INTRODUCTION TO SPECTROPOLARIMETRY

Spectropolarimetry embraces the most complete and detailed measurement and analysis of light, as well as its interaction with matter. This book provides an introductory overview of the area, which is playing an increasingly important role in modern solar observations. Chapters include a comprehensive description of the polarization state of polychromatic light and its measurement, an overview of astronomical (solar) polarimetry, the radiative transfer equation for polarized light, and the formation of spectral lines in the presence of a magnetic field. Most topics are dealt with within the realm of classical physics, although a small amount of quantum mechanics is introduced where necessary. This text will be valuable for advanced undergraduates, graduates and researchers in astrophysics, solar physics and optics.

JOSE CARLOS DEL TORO INIESTA obtained his doctorate in Physical Sciences from the Universidad de La Laguna, Spain, in 1987. He is currently a tenure scientist at the Instituto de Astrofísica de Andalucía (CSIC), Spain, where his main field of research is photospheric magnetic structures and diagnostic techniques for the analysis of polarized light. He is a member of the European Astronomical Society, the International Astronomical Union, and a senior member of the Spanish Astronomical Society.

INTRODUCTION TO SPECTROPOLARIMETRY

JOSE CARLOS DEL TORO INIESTA
Instituto de Astrofísica de Andalucía
Consejo Superior de Investigaciones Científicas

CAMBRIDGE
UNIVERSITY PRESS

CAMBRIDGE UNIVERSITY PRESS
Cambridge, New York, Melbourne, Madrid, Cape Town, Singapore, São Paulo

Cambridge University Press
The Edinburgh Building, Cambridge CB2 8RU, UK

Published in the United States of America by Cambridge University Press, New York

www.cambridge.org
Information on this title: www.cambridge.org/9780521818278

First published 2003
This digitally printed version 2007

A catalogue record for this publication is available from the British Library

Library of Congress Cataloguing in Publication data

Toro Iniesta, J. C. del (Jose Carlos)
Introduction to spectropolarimetry / Jose Carlos del Toro Iniesta.
p. cm.
Includes bibliographical references and index.
ISBN 0 521 81827 3 (hardback)
1. Astrophysical spectropolarimetry. I. Title.
QB465 .T67 2003
523.01′522–dc21 2002073454

ISBN 978-0-521-81827-8 hardback
ISBN 978-0-521-03648-1 paperback

To Enriqueta
and our joint project:
Jorge and Ana

Contents

Preface

For the object of the philosopher is not to complicate, but to simplify and analyze, so as to reduce phenomena to laws, which in their turn may be made the stepping-stones for ascending to a general theory which shall embrace them all; and when such a theory has been arrived at, and thoroughly verified, the task of deducing from it the results which ought to be observed under a combination of circumstances which has nothing to recommend it for consideration but its complexity, may well be abandoned for new and more fertile fields of research.

—G. G. Stokes, 1852.

Were one asked for a concise description of most astrophysical tasks, one possible answer might be 'understanding the message of light from heavenly bodies'. Light – or electromagnetic radiation – is the astronomer's main (almost his sole) source of information. The statement that nobody can measure the physical parameters of the solar atmosphere, although at first sight shocking, merely calls attention to the fact that astrophysics is an observational rather than an experimental science. This characteristic is often forgotten. We do not measure solar or stellar temperatures, velocities, magnetic fields, etc., simply because we do not have thermometers, tachometers, magnetometers, etc., that would permit *in situ* measurements of these parameters. Rather, we are only able to measure light. The astronomical parameters are inferred from these measurements, often with the help of some laboratory physics. Thus, the reliability of such astronomical inferences hinges on the accuracy of measurements of light. It is in this broad sense that spectropolarimetry may be said to embrace all real measurements carried out by astronomers. For, as is clear from the name itself, spectropolarimetry analyzes light as a function of its two most important characteristics: wavelength and state of polarization.

The basic laws or physical tools used in astrophysics are those already known in the laboratory, but sometimes astrophysics offers new insights on how to proceed in the laboratory. A beautiful example can be found in the subject of this book.

Spectropolarimetry was born and mostly developed in the realm of astrophysics – more specifically, in solar physics: the importance of magnetic fields in the overall state of our star has been increasingly more appreciated throughout the course of the twentieth century. Most (if not all) the observable manifestations of solar magnetism have been polarimetric. Great efforts have consequently been made in improving the accuracy of the measurements and in building a theory that permits the analysis of such spectropolarimetric measurements. Both the instrumental and the theoretical developments are of use to a much broader community. Advances in spatial and temporal modulation of the polarimetric signal achieved over the last two decades will doubtlessly be of interest for laboratory polarimetry. On the other hand, the theory of polarized radiation transport not only helps in disclosing the physical parameters of the Sun and other stars, but can be described as a cornerstone in our understanding of the interaction between radiation and matter. It is with this mutual-benefit philosophy in mind that this book has been written: though it is primarily intended for astrophysical applications, most of the concepts and developments described in the book may be of use to other branches of science. The use of astrophysical examples has been minimized as far as possible in order to give a general overview of the topics discussed. Nevertheless, most of the particular examples are still from solar physics.

Being a primordial property of the simplest (and ideal) form of light (the monochromatic, plane, time-harmonic wave), polarization is (fortunately) a valid concept and indeed a characteristic of measurable polychromatic light in the real world. Hence, understanding the message of light necessarily implies (or should imply) a polarimetric analysis of electromagnetic waves in order to exploit fully the information carried by them from heavenly bodies. The other important variable characterizing electromagnetic radiation, frequency, has received wide and intensive attention in many textbooks. Nowadays, the postgraduate student or the young researcher has many authoritative monographs available in which spectroscopy is discussed thoroughly, in both the theoretical and the observational and experimental aspects. This is not the case, however, for polarimetry, whose details are seldom discussed, with very few exceptions. After an introductory historical summary, the first part of the book (Chapters 2–5) is devoted to polarimetry and hence may be of direct application not only to solar physics but to other branches of astrophysics and science in general. But if light and its state of polarization are the primordial observables in astrophysics, the radiative transfer equation (RTE) is crucial for deciphering that encoded information in terms of the physical quantities characterizing the (polarized) light source. Therefore, the second part of the book (Chapters 6–11) deals with the RTE for polarized light with particular emphasis on the case of magnetized solar atmospheres.†

Since the present monograph was conceived as an introductory course for postgraduate or advanced undergraduate students, the reader is assumed to be familiar with such concepts as Maxwell's equations and the Fourier transform. Nevertheless, I have decided to start practically from scratch in order to lay the foundations on which the topics described are built. Therefore, a non-negligible amount of the material presented here has been borrowed from other sources [special mention should be made of the books by Born and Wolf (1993) and Stenflo (1994), and the lecture notes by Landi Degl'Innocenti (1992)].‡

Although already pointed out in several works, certain paramount details of polarimetry, such as sign conventions, the distinction between monochromatic and quasi-monochromatic light, the coherent and incoherent superposition of light beams, or the physical meaning of Mueller matrices, need particular attention within the framework of a global description. The kinship between the scalar RTE for unpolarized light and the vector RTE for polarized light (the former being a particular case of the latter) is not emphasized often enough. Likewise, deep-rooted concepts in the community, such as the *height of formation* of spectral lines often hide very important clues to the correct inference of the physical properties of the observational target and should in many cases be forgotten. These are a few reasons that justify the structure of the text.

In Chapter 2, some basic concepts are reviewed such as the description of light as an electromagnetic wave, the monochromatic time-harmonic plane wave, the polarization tensor (or coherency matrix), the Stokes parameters of a monochromatic wave, and the Poincaré sphere. Chapter 3 describes the polarization properties of quasi-monochromatic light. Polychromatic light is introduced as a statistical superposition of monochromatic light. Then, after defining a quasi-monochromatic plane wave, its associated coherency matrix and Stokes parameters are described and their meaning explained from the measurements viewpoint. Finally, the concepts of degree of polarization, natural light, partially and completely polarized light are defined. The transformations suffered by (partially) polarized light after interaction with linear optical systems are dealt with in Chapter 4, where the Mueller matrices and their properties and characterization are presented. A description of the basic block components of a (solar) polarimeter follows, and the chapter ends by outlining the way we measure, i.e., by describing both the spatial and temporal modulation of

† I use the term "atmosphere" in the plural because the Sun may be thought of as having many atmospheres, depending on whether we are studying the atmosphere above a granule, an unmagnetized intergranular region, a penumbral filament, an umbra, etc.

‡ See the references in the recommended bibliography to Chapter 1.

the polarimetric signal. Chapter 5 provides a more in-depth treatment of specific issues germane to *solar* polarimetry, such as environmental (seeing-induced and instrumental) polarization, modulation and demodulation, and a description of current solar polarimeters. Light propagation through low-density, weakly conducting media along with absorption and dispersion phenomena are studied in Chapter 6. The radiative transfer equation for polarized light is discussed in Chapter 7, and the links with the more usual scalar equation for unpolarized light are established. The symmetries and information content of the propagation ("absorption") matrix and of the source function vector in the local thermodynamic equilibrium (LTE) approximation are also discussed in this chapter. After describing the properties that the Zeeman effect imprints on spectral lines, a specification of the elements of the propagation matrix in the presence of magnetic fields and the formation of lines in a magnetized stellar atmosphere are presented in Chapter 8. The solution of the RTE follows in Chapter 9. The paramount astrophysical problem, namely, that of finding radiative transfer diagnostics that enable the astronomer to interpret the observables (the Stokes profiles) in terms of the physical parameters of the observed atmospheres is dealt with in the final two chapters. Chapter 10 describes evidence for the RTE as the most useful tool in magnetometry. First, the concept of height of formation of spectral lines is discussed critically and a recommendation is made that it be substituted by that of sensitivity of the lines to atmospheric quantities. After linearization of the RTE, the so-called response functions (RFs) are shown to describe such sensitivities fully and their main properties are outlined. Moreover, the sensitivities can be extended to observable parameters derived from the profiles, and through a theoretical generalization of the measurements, generalized RFs can be defined as well. These generalized RFs are the root of the concept of height of formation for measurements. Finally, Chapter 11 summarizes very briefly the aim and the bases of the so-called inversion techniques that are currently the best candidates for inferring the magnetic, dynamic, and thermodynamic properties of solar atmospheres.

Contrary to customary usage, just a few references will be quoted within the text. Instead, a bibliography is recommended at the end of each chapter. The number of cited research papers has been reduced to a minimum but some of these have been compiled purely because of their historical interest. By far most of the auxiliary material the reader may need can be found in books or review papers.

Granada, April 2, 2002 *Jose Carlos del Toro Iniesta*

Acknowledgements

When I was invited to give a course on 'Solar polarimetry and magnetic field measurements' at the Kanzelhöhe Observatory Summer School of 1999, I felt excited at the possibility of summarizing in a didactic manner most of the topics related to my research work of the previous 17 years. Teaching always brings significant rewards to the teacher, and I had been offered a challenging opportunity! I embarked in the project but very soon realized that the material had grown in such a way that to include all my notes in the proceedings would have been unfair to the other lecturers and contributors. That, simply, is the origin of the book you now have in front of you. I must warmly thank Arnold Hanslmeier for his invitation and many of the participants and lecturers for the interest shown and their suggestion that I write a book.

I have been working in Granada, at the Instituto de Astrofísica de Andalucía (CSIC), for the past 3 years, but most of the concepts contained in this book arose from discussions and collaborations with my friends at the Instituto de Astrofísica de Canarias. Names that come to mind are Manolo Vázquez, Fernando Moreno-Insertis, Jorge Sánchez Almeida, Javier Trujillo Bueno, Inés Rodríguez Hidalgo, Carlos Westendorp Plaza, Luis Bellot Rubio, and, most importantly, Manolo Collados, Valentín Martínez Pillet, and Basilio Ruiz Cobo. I hope I have not forgotten anybody. From all of these colleagues I have benefited in terms not only of science but also of friendship. Among the colleagues and friends from abroad, I would like to thank Bruce Lites, Andy Skumanich, Arturo López Ariste, and Héctor Socas-Navarro from the High Altitude Observatory (Boulder, Colorado), and, of course, Meir Semel from the Observatoire de Paris-Meudon and Egidio Landi Degl'Innocenti from the Università di Firenze. I learnt many things from them. Luis, Arturo, Manolo, Héctor and Basilio critically reviewed individual chapters, and Antonio Claret, of the IAA, reviewed the whole text. I am deeply indebted to them all, but, of course, any errors that may still remain are entirely my responsibility.

I must also thank Terry Mahoney for carefully checking the English and for his fine translations of the literary quotations in Spanish. Figure 5.3 has been adapted with permission from *The Observation and Analysis of Stellar Photospheres* by D. F. Gray, 2nd Edition (Cambridge University Press: Cambridge, 1992). Permissions for the opening quotations in Chapters 3 and 8 were granted by the copyright holders and are taken respectively from *Polarized light* by W. A. Shurcliff (Harvard University Press and Oxford University Press: Cambridge, Mass. & London, 1962) and from 'Modeling: an aim and a tool in astrophysics' by H. C. van de Hulst, in *Modeling the stellar environment. How and why?* (Éditions Frontières: Gif-sur-Yvette, 1989).

I have left thanking my family to the end. I cannot forget the extreme enthusiasm and support of my parents, Julián and Donita. Without their efforts I would not have been able even to enter research. My wife, Enriqueta, and our two children, Jorge and Ana, are a continuous stimulus and source of inspiration. I cannot thank them enough for their patient understanding and forbearance during the long months dedicated to the writing of this book.

J.C.T.I.

1
Historical introduction

Más has dicho, Sancho, de lo que sabes —dijo don Quijote—; que hay algunos que se cansan en saber y averiguar cosas que, después de sabidas y averiguadas, no importan un ardite al entendimiento ni a la memoria.

—*M. de Cervantes Saavedra, 1615.*

'You have said more than you realize,' said Don Quijote, 'for there are some, who exhaust themselves in learning and investigating things which, once known and verified, add not one jot to our understanding or our memory.'

Spectropolarimetry, as the name suggests, is the measurement of light that has been analyzed both spectroscopically and polarimetrically. In other words, both the wavelength distribution of energy and the vector properties of electromagnetic radiation are measured with the highest possible resolution and accuracy. Thus, spectropolarimetry embraces a number of techniques used in order to characterize light in the most exhaustive way. Such techniques are ultimately based on a theory that, from its beginnings in the closing years of the nineteenth century, finally grew to maturity in the 1990s. Therefore, under the heading of spectropolarimetry we will find several disciplines, which, despite being interrelated or rather, although our aim is to stress their interrelatedness, may be considered independent.

A historical perspective is always helpful for grasping the importance of physical phenomena and their corresponding explanations. The main objective of this chapter is to give a brief description of the salient events and findings in history related to some of the independent disciplines covered in this book. In particular, polarization phenomena and their treatment, and the Zeeman effect, both on the Sun and in the laboratory, have been especially relevant not only in spectropolarimetry but in the development of natural sciences and are thus deserving of this brief introduction.

As we shall see, polarization phenomena provided the most important observational evidence that finally helped to decide between the corpuscular and undulatory theories of light. In its turn, the Zeeman effect was not only a cornerstone in the development of quantum mechanics but also the key to the discovery and later study of extraterrestrial magnetic fields.

1.1 Early discoveries in polarization

Like many other discoveries in physics and science in general, polarization was brought to the attention of the scientific community quite by chance. A mariner returning to Copenhagen from Iceland brought back several beautiful crystals of what we now know as Iceland spar, or calcite. Some of these crystals, it seems, fell into the hands of Erasmus Bartholin, a Danish physician, mathematician, and physicist, who at that time (1669) was a professor of medicine at the University of Copenhagen. Bartholin observed that images formed through these crystals were double. Moreover, when the crystal was rotated, one image remained in place while the second rotated with the crystal. He rapidly interpreted the phenomenon in terms of rays entering the crystal being immediately split into two, one of which, being stationary during rotation, he termed the "ordinary ray", and the other – that experiencing the crystal rotation – he called the "extraordinary ray". As we shall see later, these terms are still in use.

This discovery was later taken up by the Dutch mathematician, astronomer, and physicist Christiaan Huygens, who had no problem in explaining the double refraction with his construction of propagating wavefronts (1690). At any point within the crystal, the light disturbance generated two wavefronts, a spherical one for the ordinary ray and a spheroidal one for the extraordinary ray. Besides this explanation, he contributed the important experimental discovery that the doubly diffracted rays behaved differently from ordinary light when entering a second crystal. Depending on the relative orientation of the two crystals, double refraction may or may not take place again, thus producing (or not) four final rays. He was not able, however, to provide a comprehensive explanation for this new phenomenon. Interestingly, it was Sir Isaac Newton who in 1717 presented the first ideas concerning the reason for double refraction. According to Newton, ordinary light seemed to have the same properties in all directions perpendicular to the direction of propagation, while doubly refracted rays seemed to "have sides", i.e., to show different properties in different directions in a plane perpendicular to the direction of propagation. Today, we know that this is qualitatively true. However, for Newton these peculiar transversal properties constituted an insuperable objection to accepting the wave theory of light; at that time, accepting the wave theory implied accepting similarities between light and (longitudinal) sound waves.

Most of the scientific debate concerning optics during the eighteenth century was between the rival corpuscular and undulatory theories but double refraction was still a major problem. The undulatory theory started to gain reputation by the turn of the century with the work of the English former physician Thomas Young. Apart from his great discovery of the law of interference in 1801, Young used double refraction as an argument for defending the undulatory theory. On Young's suggestion, in 1802 William Hyde Wollaston studied experimentally the accuracy of the Huygenian construction for the extraordinary ray and found remarkable agreement, although the existence itself of two rays in a single substance was still not completely understood. To get an idea of the importance of the problem suffice it to say that in January 1808 the French Academy proposed "To furnish a mathematical theory of double refraction, and to confirm it by experiment." As the subject for the 1810 physics prize. One of the contestants was Etienne-Louis Malus, an engineering officer in Napoleon's army, who happened to be analyzing solar light reflected by a window through a rhomb of Iceland spar when the two rays showed different intensities! He rightly concluded that the properties hitherto attributed to crystals could also be produced by reflection of light in a variety of substances and he communicated his results to the French Academy by the end of that year, when he coined the term *polarization*. Yet wave-theory defenders were unable to abandon the analogy with sound waves and the debate kept up for some years: including polarization phenomena within that theory was still necessary. The solution to the puzzle took 9 years, during which a number of discoveries occurred. In 1811, Dominique François Jean Arago, a French physicist, discovered optical rotation and in the following year he invented the pile-of-plates polarizer. Also in 1812, the Scottish physicist David Brewster discovered the law named after him concerning polarization by reflection, and the French physicist Jean-Baptiste Biot discovered positive and negative birefringence in uniaxial crystals.

During a visit to Young in 1816, Arago mentioned a new experiment he and Augustin-Jean Fresnel had recently carried out. The result was that two light beams polarized at right angles do not interfere as two rays of ordinary light do, and they always show the same total intensity when reunited, no matter what the path difference for the two beams. This experiment provided Young with the much-sought-after key to the link between the undulatory theory of light and polarization phenomena: the transversality of light vibrations. In 1817, first in a letter to Arago and then in an article for *Encyclopaedia Britannica*, he explained that the assumption of light oscillations perpendicular to the propagation direction fully accounts for the "subdivision of polarised light by reflection in an oblique plane". Young's ideas, communicated by Arago to Fresnel in 1819, were quickly realized by the latter as the main explanation for their experimental results and in fact for all the polarization phenomena known until then. Fresnel rapidly interpreted natural light as a superposition of light polarized in all possible directions and polarization as a manifestation of wave transversality:

So direct light can be considered as the union, or more exactly as the rapid succession, of systems of waves polarised in all directions. According to this way of looking at the matter, the act of polarisation consists not in creating these transverse motions, but in decomposing them into invariable directions, and separating the components from each other; for then, in each of them, the oscillatory motions take place always in the same plane.

These results on polarization laid the foundation for Fresnel's further important discoveries in optics.

1.2 A mathematical formulation of polarization

In his remarkable paper of 1852 entitled "On the composition and resolution of streams of polarized light from different sources", George Gabriel Stokes, Lucasian Professor of Mathematics at the University of Cambridge, established a mathematical formalism ideal for describing the state of polarization of any light beam. Moreover he demonstrated several of the most important properties of polarized light, among which he noted the following:

When any number of independent polarized streams, of given refrangibility, are mixed together, the nature of the mixture is completely determined by the values of four constants, which are certain functions of the intensities of the streams, and of the azimuths and eccentricities of the ellipses by which they are respectively characterized; so that any two groups of polarized streams which furnish the same values for each of these four constants are optically equivalent.

Those four constants referred to by Stokes are what we currently know as Stokes parameters. Unfortunately, the usefulness of the formalism and the importance of the Stokes theorems seems to have been ignored by the scientific community during the following 80 years. In 1929, in a very complete study of the partial polarization of light, the French physicist Paul Soleillet described the Stokes parameters and used them throughout. Interestingly, in the third part of this very paper, a formulation is presented of a theory of anisotropic absorption that is nothing less than the construction of an equation of transfer for polarized radiation. Unfortunately, this paper is still fairly unknown by the astrophysical community. Eighteen years later, in 1947, in his famous series of papers on the radiative equilibrium of stellar atmospheres, and certainly unaware of Soleillet's work, the Indian-born American astrophysicist Subrahmanyan Chandrasekhar published a summary of Stokes's results, emphasizing the importance and usefulness of the formalism which turned out to be particularly well suited to the formulation of a radiative transfer equation in a stellar atmosphere.

One year later, in 1948, Hans Mueller, a professor of physics at the Massachusetts Institute of Technology devised a phenomenological approach to describing the transformation of Stokes parameters by means of 4×4 matrices (nowadays known as Mueller matrices). Since then, Mueller's approach has been extensively used for dealing with partially polarized light. A precursor of this formalism can be found in a paper by Francis Perrin (1942). A few years before Mueller's work, between 1941 and 1947, the American physicist Robert Clark Jones presented his formalism to describe totally polarized light and the transformations between any two totally polarized light beams.

1.3 Discovery of the Zeeman effect

When spectral lines are formed in the presence of a magnetic field, they widen or split into differently polarized components. This phenomenon is known as the Zeeman effect, in honor of its discoverer, the Dutch physicist Pieter Zeeman, who in 1896 found a conspicuous widening of the sodium D lines after switching on an electromagnet. But the origins and precursors of this discovery date back to the middle of the nineteenth century, as does the observation of this particular effect over the surface of the Sun.

In 1845, Michael Faraday discovered that linearly polarized light streaming through a transparent isotropic medium subject to a magnetic field changes the direction of polarization. This relationship between magnetism and light was his inspiration for his final scientific endeavors in 1862 in searching for any trace of the influence of magnetic fields in the spectra of several substances. Unfortunately, he failed to obtain any experimental evidence. Neither did Peter Guthrie Tait of the University of Edinburgh who in 1875, influenced by the mechanical analogies of electromagnetism of William Thomson (Lord Kelvin), had a similar intuition to Faraday's. Ten years later, the Belgian astronomer M. Fievez carried out laboratory experiments in which he did find some indications of a magnetic influence on the sodium spectral lines. He was not able to discriminate magnetic from temperature effects, so he stopped his inquiries at that point.

Unacquainted with this work, Pieter Zeeman, associate professor at the University of Leyden, had the same intuition as Faraday and Tait because of his work on the Kerr effect. His first experiments failed to find any observable effect, and he would have not tried again had he not read by chance, in the *Collected Works* of James Clerk Maxwell, of the final efforts of Faraday. Having had ideas similar to Faraday's encouraged him to take up the experiments with more care. His experimental results were soon forthcoming, and explanations for them came from Hendrik Antoon Lorentz, also a professor at Leyden University. The widening of spectral lines had to be accompanied by a distinct polarization in the wings of the lines. Zeeman and Lorentz shared the Nobel Prize for Physics in 1902. Yet disagreements between experiment and theory were very quickly found, and the Zeeman effect remained unexplained for some 30 years after its discovery. Hence, it constituted an experimental milestone in the development of quantum mechanics. Only after the electron theories of Wolfgang Ernst Pauli (1927; non-relativistic) and Paul Adrien Maurice Dirac (1928; relativistic) could the empirical results be fully understood.

In parallel with laboratory discoveries, astronomical spectroscopy received a great impetus from the middle of the century. In 1866, Sir Joseph Norman Lockyer observed the spectrum of a sunspot. Comparison with the spectrum of the normal solar photosphere revealed a conspicuous widening of the lines. This phenomenon, interpreted nowadays as a result of the Zeeman effect brought about by the sunspot's magnetic field, was observed by many workers until the early 1900s. Unfortunately, nobody realized the relevance of the phenomenon, even well after Zeeman's discovery had been brought to the notice of the astrophysical community.

Motivated by the new morphology of sunspots as seen in H_α photographs, in 1908 George Ellery Hale finally found a convincing explanation for the observed sunspot spectrum: the presence of strong magnetic fields in sunspots. The spectral lines appeared to be widened, split, and in the right state of polarization. This can be thought of as one of the fundamental discoveries of solar physics in the twentieth century. It triggered new and fertile fields of solar and stellar research, for which spectroscopy and polarimetry must be combined in order to exploit to the full the information embedded in the light from heavenly bodies.

1.4 Radiative transfer for polarized light

The specification of the radiation field through an atmosphere that scatters light started as a physical problem in 1871 with the work of the English physicist John William Strutt (Lord Rayleigh) on sky light. Independently of the already-mentioned paper of Soleillet but a few years later, the fundamental equations were formulated and solved by Subrahmanyan Chandrasekhar in his series of papers published in *The Astrophysical Journal*. In the meantime, the works of Arthur Schuster (1905) and Karl Schwarzschild (1906) deserve especial mention because of their revival of the problem mainly within the astrophysical community. By the middle of the twentieth century, however, physicists from other branches became interested in radiative transfer since the same problems seemed to arise in, for instance, the diffusion of neutrons. Most remarkably, the transfer was formulated by Chandrasekhar for polarized radiation since the original problems had to do with light polarized by scattering. Nevertheless, since the 1940s the most extended use of the radiative transfer equation has been in relation to unpolarized problems:

stellar atmospheres have often been assumed isotropic so that just one equation for the total intensity of the light beam is needed.

The study of solar and stellar magnetic fields is an application for which the problem of polarized energy transport is of singular importance. The wealth of information obtained after Hale's discovery of sunspot magnetic fields made it necessary to interpret the spectrum of polarized light observed in the Sun and other stars. Since the mid-1950s, a full theory of polarized radiative transfer has been developed, mostly motivated by the problem of solar/stellar magnetic fields, but whose applications go far beyond astrophysics. Since the theory is relatively young, it would seem appropriate to mention a few landmarks in the literature. The theory builds upon the pioneering work by Wasaburo Unno from Japan (1956) dealing with the formulation – and solution in a simplified Milne–Eddington model – of a radiative transfer equation in the presence of a magnetic field. The work was indeed aimed at describing spectral line formation in the presence of magnetic fields in the solar atmosphere. Only absorption processes were taken into account and the completion of such an equation, including dispersion effects (the so-called magneto-optical effects in the astrophysical literature) was carried out by D. N. Rachkovsky (1962a, 1962b, 1967) from the Ukraine – of course, the solution in that simplified model was also corrected. The formulation, however, was phenomenological and somewhat heuristic and was not put on a firm, rigorous basis until the work by Egidio Landi Degl'Innocenti and Maurizio Landi Degl'Innocenti (1972), who derived the transfer equation for polarized light within the framework of quantum electrodynamics. Later, three derivations of that equation from first principles (basically Maxwell's equations) of classical physics were published: one by John Jefferies, Bruce W. Lites, and Andrew P. Skumanich (1989), another by Jan Olof Stenflo (1991; see also 1994), and a third by Egidio Landi Degl'Innocenti (1992). Many of the developments which follow in this text are based on these three works.

Recommended bibliography

Born, M. and Wolf, E. (1993). *Principles of optics*, 6th edition (Pergamon Press: Oxford). Historical introduction.

Chandrasekhar, S. (1946). On the radiative equilibrium of a stellar atmosphere. X. *Astrophys. J.* **103**, 351.

Chandrasekhar, S. (1946). On the radiative equilibrium of a stellar atmosphere. XI. *Astrophys. J.* **104**, 110.

Chandrasekhar, S. (1947). On the radiative equilibrium of a stellar atmosphere. XV. *Astrophys. J.* **105**, 424.

Chandrasekhar, S. (1960). *Radiative transfer* (Dover Publications, Inc.: New York). Chapter 1. (See bibliography for references of Schuster and Schwarzschild.)

Hale, G.E. (1908). On the probable existence of a magnetic field in Sun-spots. *Astrophys. J.* **28**, 315.

Jefferies, J., Lites, B.W., and Skumanich, A.P. (1989). Transfer of line radiation in a magnetic field. *Astrophys. J.* **343**, 920.

Jones, R.C. (1941). New calculus for the treatment of optical systems. I. Description and discussion of the calculus. *J. Opt. Soc. Am.* **31**, 488. (See Shurcliff below for subsequent papers.)

Landi Degl'Innocenti, E. (1992). Magnetic field measurements, in *Solar observations: techniques and interpretations*. F. Sánchez, M. Collados, and M. Vázquez (eds.) (Cambridge University Press: Cambridge). Sections 4–8.

Landi Degl'Innocenti, E. and Landi Degl'Innocenti, M. (1972). Quantum theory of line formation in a magnetic field. *Solar Phys.* **27**, 319.

Lockyer, J.N. (1866). Spectroscopic observations of the Sun. *Proc. Roy. Soc.* **15**, 256.

Mueller, H. (1948). The foundation of optics. *J. Opt. Soc. Am.* **38**, 661.

Perrin, F. (1942). Polarization of light scattered by isotropic opalescent media. *J. Chem. Phys.* **10**, 415.

Rachkovsky, D.N. (1962a). Magneto-optical effects in spectral lines of sunspots (in Russian). *Izv. Krymsk. Astrofiz. Obs.* **27**, 148.

Rachkovsky, D.N. (1962b). Magnetic rotation effects in spectral lines (in Russian). *Izv. Krymsk. Astrofiz. Obs.* **28**, 259.

Rachkovsky, D.N. (1967). The reduction for anomalous dispersion in the theory of the absorption line formation in a magnetic field (in Russian). *Izv. Krymsk. Astrofiz. Obs.* **37**, 56.

Shurcliff, W.A. (1962). *Polarized light* (Harvard University Press & Oxford University Press: Cambridge, MA & London). Chapter 1.

Soleillet, P. (1929). Sur les paramètres caractérisant la polarisation partielle de la lumière dans les phénomènes de fluorescence. *Ann. Phys.* **12**, 23.

Stenflo, J.O. (1991). Unified classical theory of line formation in a magnetic field, in *Solar polarimetry*. L.J. November (ed.) (National Solar Observatory Sacramento Peak: Sunspot, NM), 416.

Stenflo, J.O. (1994). *Solar magnetic fields. Polarized radiation diagnostics* (Kluwer Academic Publishers: Dordrecht). Chapter 3.

Stokes, G.G. (1852). On the composition and resolution of streams of polarized light from different sources. *Trans. Cambridge Phil. Soc.* **9**, 399.

del Toro Iniesta, J.C. (1996). On the discovery of the Zeeman effect on the Sun and in the laboratory. *Vistas in Astronomy* **40**, 241.

Unno, W. (1956). Line formation of a normal Zeeman triplet. *Publ. Astron. Soc. Japan* **8**, 108.

Whittaker, E. (1973). *A history of the theories of aether and electricity*, 1st Vol. (Humanities Press: New York). Chapters 1, 4, and 13.

Zeeman, P. (1897). On the influence of magnetism on the nature of the light emitted by a substance. *Astrophys. J.* **5**, 332. See also *Phil. Mag.* **43**, 226.

2

A review of some basic concepts

Finalmente, quiero, Sancho, me digas lo que acerca desto ha llegado a tus oídos; y esto me has de decir sin añadir al bien ni quitar al mal cosa alguna, . . .

—*M. de Cervantes Saavedra, 1615.*

'Finally, Sancho, I want you to tell me what has reached your ears concerning this matter, and you must do so without adorning the good or lessening the ill.'

This chapter is devoted to recalling a number of results of importance for development in later chapters. Most of these concepts are assumed to be already known to the reader, and those derivations that are missing will be found in textbooks on optics and electromagnetism. The main aim here is to provide a summary of the polarization properties of the simplest electromagnetic wave one can conceive: the monochromatic, time-harmonic, plane wave.

The terms *light* and *electromagnetic wave* will be understood as synonymous throughout the text. More specifically, we will be referring to the visible part of the spectrum and its two nearest neighbors, the ultraviolet and the infrared. Many of the topics discussed are also applicable to other wavelength regions. In particular, it is worth noting that radio observations use most of the concepts we shall be developing here for the optical region, although they are not in principle necessary for that wavelength range. The relevant difference between our spectral region and those of radio waves on the one hand and X- and γ rays on the other lies in the means of detection or measurement. These techniques are necessarily different owing to the basically different characteristic frequencies of electromagnetic waves in the three regimes. Radio waves have a sufficiently small frequency for the amplitude and phase of the oscillating electric field to be measured directly. The oscillation is so slow that it can be "followed" by antennas whose response is proportional to the electric field. In the optical regime (including the infrared and the ultraviolet), the frequencies are so high that the electric oscillations cannot be followed by any detector. The response of detectors is proportional to the electromagnetic energy carried by the waves. Finally, in the most energetic domain, the interaction between radiation and matter is often better characterized by the corpuscular rather than by the undulatory properties of radiation. Treatment of radio, X-, and γ rays is beyond the scope of this book.

2.1 Light as an electromagnetic wave

In regions free of currents and charges in a homogeneous isotropic medium, Maxwell's equations imply that the electric field vector, $\boldsymbol{E}$, associated with the luminous perturbation must verify the homogeneous wave equation

$$\nabla^2 \boldsymbol{E} - \frac{\varepsilon\mu}{c^2}\ddot{\boldsymbol{E}} = \boldsymbol{0}, \tag{2.1}$$

where ε stands for the dielectric permittivity, μ for the magnetic permeability of the medium, and c for the speed of light *in vacuo*. An analogous equation holds for the magnetic field vector, $\boldsymbol{H}$.

By *isotropic* one means a medium whose behavior under the influence of the electromagnetic field is well characterized by material equations with scalar coefficients ε, μ, and σ (the conductivity). Hence, the material behavior at every point is independent of the propagation direction of the field. If, besides, $\nabla(\ln\varepsilon) = \nabla(\ln\mu) = \boldsymbol{0}$, i.e., if no directional variations of ε and μ exist, the medium is said to be *homogeneous*.

The coefficient in Eq. (2.1) can be readily interpreted as the velocity of electromagnetic waves propagating through the medium:

$$v = \frac{c}{\sqrt{\varepsilon\mu}} \equiv \frac{c}{n}, \tag{2.2}$$

where the new quantity n is called the refractive index of the medium. By interpreting $c/\sqrt{\varepsilon\mu}$ as a propagation velocity, we are in fact neglecting (possible) standing-wave solutions to Eq. (2.1). The isotropic medium is characterized by a single refractive index and the electromagnetic waves propagate throughout with a single velocity no matter what the direction.*

The simplest solution of Eq. (2.1) is that of a plane wave for which each Cartesian component of $\boldsymbol{E}$ (and $\boldsymbol{H}$), at a given point, $\boldsymbol{r}$, in space and at a time t, depends only on the magnitude $u \equiv \boldsymbol{r}\cdot\hat{\boldsymbol{s}} - vt$ of the vector $\boldsymbol{u} = u\hat{\boldsymbol{s}}$:

$$\boldsymbol{E}(\boldsymbol{r}, \hat{\boldsymbol{s}}, t) = \boldsymbol{E}(\boldsymbol{r}\cdot\hat{\boldsymbol{s}} - vt); \quad \boldsymbol{H}(\boldsymbol{r}, \hat{\boldsymbol{s}}, t) = \boldsymbol{H}(\boldsymbol{r}\cdot\hat{\boldsymbol{s}} - vt), \tag{2.3}$$

* Although very simple and well known to the reader, this digression about homogeneity and isotropy of the medium is in order. Many of the discussions concerning polarimetry and polarized radiative transfer are based on anisotropies and inhomogeneities of the medium through which the light is traveling.

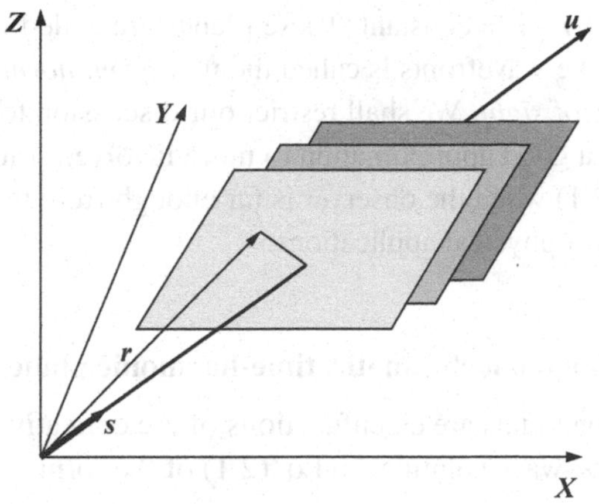

Fig. 2.1. A plane wave depends only on $u = \boldsymbol{r} \cdot \hat{s} - vt$. Then, the electric field is constant over planes perpendicular to the propagation direction at a given instant t.

where $\hat{s}$ is a unit vector in the direction of propagation and the dot symbol indicates the scalar product.

This wave can be shown (e.g., Born and Wolf, 1993) to verify that

$$\boldsymbol{E} = -\sqrt{\frac{\mu}{\varepsilon}}\hat{s} \wedge \boldsymbol{H} = -\frac{\mu}{n}\hat{s} \wedge \boldsymbol{H}, \qquad \boldsymbol{H} = \frac{\varepsilon}{n}\hat{s} \wedge \boldsymbol{E}, \tag{2.4}$$

where the symbol $\wedge$ represents the vector product. These equations clearly indicate that $\boldsymbol{H}$ is known whenever we know $\boldsymbol{E}$ and vice versa. On the one hand, this result allows us to neglect the equation for $\boldsymbol{H}$ homologous to Eq. (2.1). We can then study the electromagnetic wave by simply considering solutions to the electric-field wave equation (2.1). On the other hand, scalar multiplication of Eqs (2.4) by $\hat{s}$ yields

$$\boldsymbol{E} \cdot \hat{s} = \boldsymbol{H} \cdot \hat{s} = 0; \tag{2.5}$$

that is, electromagnetic waves are *transversal*, such that $\boldsymbol{E}$ and $\boldsymbol{H}$ always remain in planes perpendicular to $\hat{s}$. Finally, Eqs (2.4) and (2.5) explain that $\boldsymbol{E}$, $\boldsymbol{H}$, and $\hat{s}$ form a right-handed orthogonal triad, and that

$$\sqrt{\mu}H = \sqrt{\varepsilon}E, \tag{2.6}$$

so that, when $\varepsilon = \mu = 1$, as *in vacuo*, the electric and magnetic fields associated with the luminous perturbation have the same magnitude.

An illustration of plane waves can be found in Fig. 2.1, where the reason for the name "plane waves" is easily understood: at a given instant t, $\boldsymbol{E}$ (and $\boldsymbol{H}$) is constant over planes given by $\boldsymbol{r} \cdot \hat{\boldsymbol{s}} =$ constant. These planes are called *wavefronts*. The direction $\hat{\boldsymbol{s}}$ normal to the wavefronts is called the *wavefront normal*. In astrophysics, $-\hat{\boldsymbol{s}}$ is called the *line of sight*. We shall restrict our discussion to plane waves since these turn out to be a good approximation to more involved solutions (e.g., spherical waves) of Eq. (2.1) when the observer is far enough from the light source, as is the case for most astrophysical applications.

2.2 The monochromatic, time-harmonic plane wave

Since complex exponentials are eigenfunctions of the derivative operator, it is natural to look for plane-wave solutions to Eq. (2.1) of the form

$$E_j = a_j \, \mathrm{e}^{\mathrm{i}(ku+\delta_j)}, \tag{2.7}$$

where the index j applies to the x, y, and z Cartesian components of vector $\boldsymbol{E}$; a_j stands for the constant (real) amplitude of each component, δ_j is a constant phase factor to account for time lags between the components, k is the so-called *wave number* and is a dimensional constant to make the exponential argument dimensionless,

$$u = \boldsymbol{r} \cdot \hat{\boldsymbol{s}} - vt \equiv \frac{1}{k}(\boldsymbol{k} \cdot \boldsymbol{r} - \omega t), \tag{2.8}$$

where, by definition, $\omega = 2\pi \nu$ is the angular frequency, $\boldsymbol{k} \equiv k\hat{\boldsymbol{s}} = (2\pi/\lambda)\hat{\boldsymbol{s}}$ is the so-called wave vector, λ and ν being the wavelength and frequency of the plane wave. Hence, Eq. (2.7) can be recast in the form

$$E_j = A_j \, \mathrm{e}^{-\mathrm{i}(\omega t-\delta_j)}, \tag{2.9}$$

where A_j is the *complex* amplitude of the wave ($A_j = a_j \, \mathrm{e}^{\mathrm{i}\boldsymbol{k}\cdot\boldsymbol{r}}$). Accounting for Eq. (2.5), Eq. (2.9) gives explicitly:

$$\begin{aligned} E_x &= A_x \, \mathrm{e}^{-\mathrm{i}(\omega t-\delta_x)}, \\ E_y &= A_y \, \mathrm{e}^{-\mathrm{i}(\omega t-\delta_y)}, \\ E_z &= 0, \end{aligned} \tag{2.10}$$

if $\hat{\boldsymbol{s}} = \hat{\boldsymbol{z}}$, i.e., if light propagates along the positive Z axis. Solutions (2.10) of the wave equation (2.1) are called *monochromatic* (of a single color or frequency), *time-harmonic* (sinusoidally periodic with time), *plane waves*. We shall hereafter refer to them as just monochromatic waves.

Our choice of a complex representation for the electric and magnetic fields associated with radiation is convenient for the calculus. It is important to remark, however, that the relevant physical quantities are the real parts of these complex fields: measurable quantities must be real, and complex quantities are just a useful mathematical representation and cannot be measured. The usefulness of complex notation is somehow counterbalanced by the added difficulty that arises because of sign conventions. Equation (2.7) could have been written with a minus sign in the exponential argument as well, since the real part of E_j would be the same. In that case, the new sign convention should be preserved whenever necessary in all ensuing transformations in order to get the right results: it is the price to be paid for using the otherwise convenient complex representation. As gently pointed out by Rees (1987) and others, sign conventions are very relevant in polarimetry and should always be borne in mind.

2.3 The polarization tensor

One of the direct consequences of Maxwell's equations is the identification of the energy density of the electromagnetic field and its energy flux across a unit surface perpendicular to the propagation direction (e.g., Born and Wolf, 1993). These two quantities have their respective counterparts in the astrophysical context in terms of specific intensity and flux (see Mihalas, 1978). The *volume energy density* is given by

$$w \equiv \frac{1}{16\pi}(\varepsilon \boldsymbol{E} \cdot \boldsymbol{E}^* + \mu \boldsymbol{H} \cdot \boldsymbol{H}^*), \tag{2.11}$$

or, according to Eq. (2.6),

$$w = \frac{\varepsilon}{8\pi} E E^*. \tag{2.12}$$

The *energy flux density* is given by the *Poynting vector*,

$$\boldsymbol{S} \equiv \frac{c}{8\pi} \boldsymbol{E} \wedge \boldsymbol{H}^* = \frac{v\varepsilon}{8\pi} E E^* \hat{\boldsymbol{s}}, \tag{2.13}$$

where the asterisk stands for the complex conjugate. These definitions of the energy density and of the Poynting vector are convenient for the average process carried out in practical measurements (Sections 3.3 and 3.5) and correspond to the quantities averaged over a period of the wave.

From an observational viewpoint, one is mainly interested in these energy quantities, that is, in those quantities concerning the electromagnetic wave that are measurable with available devices. As noted in the introduction to this chapter, the electric field associated with visible light cannot be measured. Detectors are sensitive to electromagnetic energy. Therefore, Eqs. (2.12) and (2.13) lead us directly to consider the so-called *polarization tensor* or *coherency matrix* of a monochromatic wave:

$$\mathbf{C} \equiv \begin{pmatrix} E_x E_x^* & E_x E_y^* \\ E_y E_x^* & E_y E_y^* \end{pmatrix}. \tag{2.14}$$

As we shall see, the coherency matrix characterizes both the energy content and the vectorial properties of the electromagnetic wave.

With the help of Eq. (2.9) we can identify the coherency matrix elements and rewrite $\mathbf{C}$ as

$$\mathbf{C} = \begin{pmatrix} a_x^2 & a_x a_y \, \mathrm{e}^{i\delta} \\ a_x a_y \, \mathrm{e}^{-i\delta} & a_y^2 \end{pmatrix}, \tag{2.15}$$

where

$$\delta \equiv \delta_x - \delta_y \tag{2.16}$$

is the *constant* phase difference between the x and y components of vector $\boldsymbol{E}$. Note that only the *real* amplitudes appear in Eq. (2.15): the complex spatial exponentials cancel out; hence, the polarization tensor (and the energy flow) depends neither on time nor on space in the absence of sources and sinks. The diagonal elements are squares of amplitudes, i.e., intensities (or energies) except for a constant factor. They are the terms needed to evaluate both w and $\mathbf{S}$. If one considers the electromagnetic wave as the superposition of two waves, one with the electric vector oscillating along the X axis and the other with the electric vector oscillating along the Y axis, a_x^2 is proportional to the intensity of the first wave and a_y^2 is proportional to the intensity of the second wave. The sum of these diagonal terms is then proportional to the total intensity of the wave. Thus, they are directly measurable. The non-diagonal terms are complex conjugates of each other and also have the same dimensions as energy. They describe the phase relationship or correlation between the two independent Cartesian-component waves. They are not measurable because they are complex quantities but we are able to measure *real* linear combinations of them (Sections 2.4, 3.3, and 3.5).

In summary, the monochromatic wave is such that its polarization tensor has a trace

$$\mathrm{Tr}(\mathbf{C}) = a_x^2 + a_y^2 \geq 0, \tag{2.17}$$

which is proportional to the total intensity of the wave, and a zero determinant:

$$\det(\mathbf{C}) = 0. \tag{2.18}$$

Inequality (2.17) holds for every plane wave, monochromatic or otherwise (Section 3.3). The physical reason is very simple: the total intensity cannot be negative. This is not the case for condition (2.18) however. Polychromatic light may

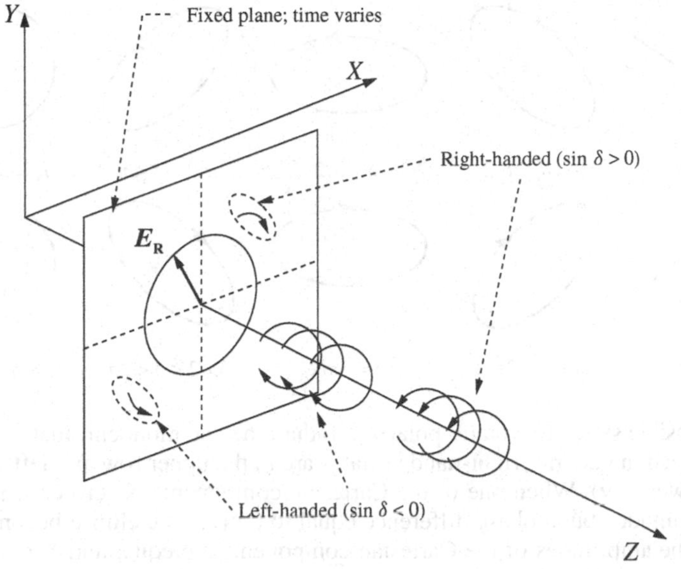

Fig. 2.2. The real part of the electric field vector rotates clockwise as seen by the observer in fixed planes as time varies if light is right-handed elliptically polarized. If the time is fixed then E_R describes the traces of a right-handed screw without rotation. When light is left-handed elliptically polarized, the opposite motions can be observed.

or may not verify it. Light with det(**C**) = 0 is said to be *totally* (or *completely*) *polarized*. Therefore, our paradigmatic, monochromatic wave is totally polarized. Total polarization means that the associated electric field vector undergoes a given motion with time in every fixed plane perpendicular to the propagation direction. Such a motion is of an ellipse given by

$$\frac{E_{x,\mathrm{R}}^2}{a_x^2} + \frac{E_{y,\mathrm{R}}^2}{a_y^2} - 2\frac{E_{x,\mathrm{R}}}{a_x}\frac{E_{y,\mathrm{R}}}{a_y}\cos\delta = \sin^2\delta, \tag{2.19}$$

as can easily be deduced from Eqs (2.10) and (2.16). In Eq. (2.19), $E_{x,\mathrm{R}}$ and $E_{y,\mathrm{R}}$ stand for the real parts of E_x and E_y, respectively. Figure 2.2 illustrates the elliptical motion with time of the real part of the electric field vector in any fixed plane perpendicular to the line of sight. There are two possible senses of rotation of the elliptically polarized field depending on the sign of $\sin\delta$. When $\sin\delta > 0$, $\boldsymbol{E}_\mathrm{R}$ undergoes a clockwise rotation as seen by the observer. The light is then said to be *right-handed elliptically polarized*. If $\sin\delta < 0$, the rotation is seen as counter-clockwise by the observer. The light is then said to be *left-handed elliptically polarized*. Should we instead have chosen a fixed instant of time, the tip of $\boldsymbol{E}_\mathrm{R}$ would have described the traces of a right-handed screw ($\sin\delta > 0$) or a left-handed screw ($\sin\delta < 0$), without rotation along the positive Z axis. All

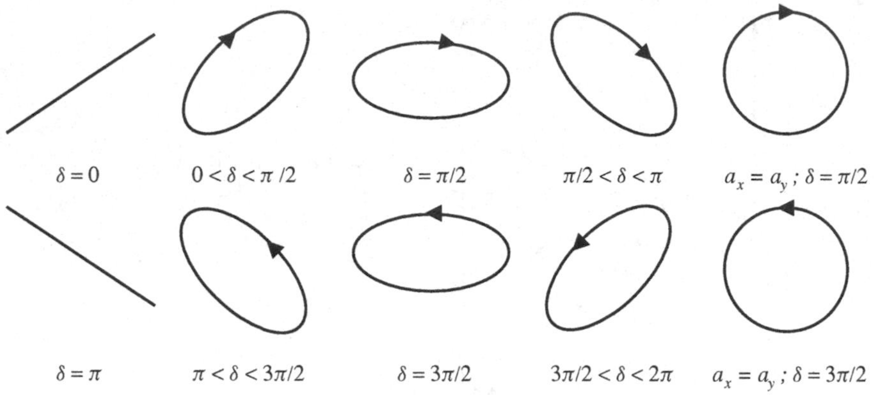

Fig. 2.3. Possible states for totally polarized radiation. The monochromatic wave is elliptically polarized in general (right-handed states are in the upper row and left-handed states are in the lower row). When one of the Cartesian components is zero or when both have different amplitudes but a phase difference equal to 0 or π, the ellipse becomes a straight line. When the amplitudes of the Cartesian components are equal and $\delta = \pi/2$ or $3\pi/2$, the ellipse becomes a circle.

possible states of polarization of a monochromatic wave are sketched in Fig. 2.3, where one can clearly see that when either $\delta = 0$ or $\delta = \pi$ the ellipse becomes a straight line. The monochromatic wave is then said to be *linearly polarized.* In the particular case when $a_x = a_y$ and $\delta = \pi/2$ or $3\pi/2$ the ellipse degenerates into a circle and the light is said to be *circularly polarized.* Note the sign convention we are adopting. One can never over-emphasize the importance of keeping in mind the arbitrary conventions. Results are of course independent of conventions but one may easily be caught out in intermediate calculations by simple sign mistakes.

2.4 The Stokes parameters of a monochromatic, time-harmonic plane wave

As far as polarization is concerned, the characterization of a monochromatic wave depends only upon *three* independent parameters; namely, the two *real* amplitudes of the x and y components and the phase difference between them. This is the physical reason for the existence of the binding condition (2.18) among the *four* matrix elements of the polarization tensor: four elements *minus* one condition give three degrees of freedom.

Our interest in the coherency matrix stems from its relation to the energy flow. The non-diagonal elements of **C**, however, are complex and are therefore not measurable: we only know how to measure real quantities. Therefore, rather than using these four matrix elements of **C**, it is customary to consider four *real* linear combinations of them: the so-called Stokes parameters which are given by

$$\begin{aligned} I &\equiv \kappa(C_{11}+C_{22}) = \kappa(a_x^2+a_y^2), \\ Q &\equiv \kappa(C_{11}-C_{22}) = \kappa(a_x^2-a_y^2), \\ U &\equiv \kappa(C_{12}+C_{21}) = 2\kappa a_x a_y \cos\delta, \\ V &\equiv i\kappa(C_{21}-C_{12}) = 2\kappa a_x a_y \sin\delta, \end{aligned} \tag{2.20}$$

where κ is a dimensional constant that translates the Stokes parameters into intensity units.† Stokes V can also be found in other texts with a minus sign (e.g., Landi Degl'Innocenti, 1992). Such a definition obeys another choice in the arbitrary convention used to define the coherency matrix. These authors define **C** as the complex conjugate of ours. The actual value of V is, however, invariant if the signs are handled with care because V is always real.

Like C_{ij} elements, the four Stokes parameters are not independent but are subject to two conditions. In order to find these conditions, it is convenient to bear in mind that, equivalently to Eq. (2.20), one can write

$$\mathbf{C} = \frac{1}{2\kappa}\begin{pmatrix} I+Q & U+iV \\ U-iV & I-Q \end{pmatrix}. \tag{2.21}$$

From this equation it is easy to see that the binding condition (2.18) becomes

$$I^2 - Q^2 - U^2 - V^2 = 0. \tag{2.22}$$

Since Stokes I represents the total intensity of the beam ($I \propto \mathrm{Tr}[\mathbf{C}]$), physical meaning [inequality (2.17)] demands that

$$I \geq 0. \tag{2.23}$$

Therefore, the Stokes parameters of the monochromatic wave are such that the first must always be positive and the sum of the squares of the last three parameters equals the square of the first.

Disregarding the case of $I = 0$ (no light), Eq. (2.22) can always be put in the form

$$\frac{Q^2}{I^2} + \frac{U^2}{I^2} + \frac{V^2}{I^2} = 1, \tag{2.24}$$

which is the equation of the surface of a sphere of radius unity, the so-called

† I do not enter into details of measurability because, in fact, monochromatic waves cannot exist except as idealized mathematical entities. Note that they should be infinite both in time and space to be purely monochromatic whereas only finite time or space intervals are available in measurements. I have introduced the Stokes parameters of a monochromatic light beam both by tradition and because of its conceptual ease. Measurability of the Stokes parameters of polychromatic light will be discussed in Sections 3.3 and 3.5.

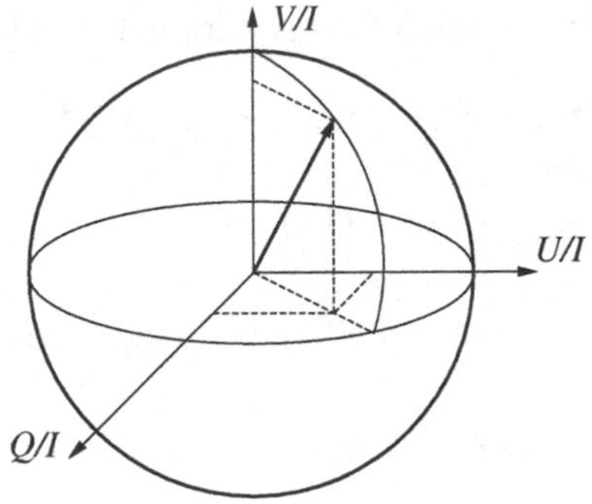

Fig. 2.4. The Poincaré sphere has radius unity. Each point on its surface represents a totally polarized state of light.

Poincaré sphere, $\mathbb{P}$ (see Fig. 2.4). Every point on the surface of such a sphere represents one of the possible polarization states of a monochromatic beam of light.

Recommended bibliography

Born, M. and Wolf, E. (1993). *Principles of optics*, 6th edition (Pergamon Press: Oxford). Chapter 1. Sections 1.1–1.4.

Chandrasekhar, S. (1960). *Radiative transfer* (Dover Publications, Inc.: New York). Chapter 1. Section 15.

Jackson, J.D. (1975). *Classical electrodynamics*, 2nd edition (John Wiley & Sons, Inc.: New York). Chapter 1. Chapter 6. Sections 6.3–6.9. Chapter 7. Sections 7.1–7.2.

Landi Degl'Innocenti, E. (1992). Magnetic field measurements, in *Solar observations: techniques and interpretations*. F. Sánchez, M. Collados, and M. Vázquez (eds.) (Cambridge University Press: Cambridge). Section 1.

Mihalas, D. (1978). *Stellar atmospheres*, 2nd edition (W.H. Freeman and Company: San Francisco). Chapter 1.

Rees, D.E. (1987). A gentle introduction to polarized radiative transfer, in *Numerical radiative transfer*. W. Kalkofen (ed.) (Cambridge University Press: Cambridge). Sections 1–2.

Shurcliff, W.A. (1962). *Polarized light* (Harvard University Press & Oxford University Press: Cambridge, Mass. & London). Chapters 1 and 2.

Stokes, G.G. (1852). On the composition and resolution of streams of polarized light from different sources. *Trans. Cambridge Phil. Soc.* **9**, 399.

3

The polarization properties of quasi-monochromatic light

If light is man's most useful tool, polarized light is the quintessence of utility.

—*W. A. Shurcliff, 1962.*

So far, the polarization properties of the simplest conceivable electromagnetic radiation have been described. However, building a polarization theory that is useful in the real world necessarily requires the consideration of light whose spectrum contains a continuous distribution of monochromatic plane waves within a finite width of frequencies. Heisenberg's uncertainty principle implies infinite time intervals for detecting purely monochromatic light (in other words, we can simply say that monochromatic light does not exist in reality). In this section we shall see that the concept of polarization is also applicable to polychromatic light. As a matter of fact, polychromatic light may share the properties of totally polarized radiation and hence be indistinguishable from monochromatic light in so far as polarimetric measurements are concerned. The coherency matrix and the Stokes parameters can also be defined for a polychromatic light beam, although the binding conditions (2.18) for **C** and (2.22) for I, Q, U, and V will be slightly modified and the new concepts of partial polarization and degree of polarization will naturally come into play.

3.1 Polychromatic light as a statistical superposition of monochromatic light

Under the hypotheses of linearity, stationarity, and continuity, one can assume any polychromatic light beam to be the superposition of monochromatic, time-harmonic plane waves of different frequencies within an interval of width $\Delta\nu$ around a central frequency ν_0. Such a superposition can be represented mathematically by writing each of the electric field Cartesian components as a Fourier

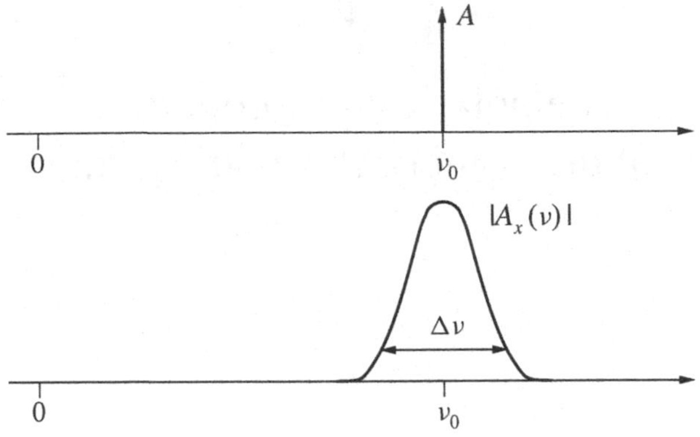

Fig. 3.1. Amplitude spectrum of a monochromatic wave (top row). Amplitude spectrum of a polychromatic wave (bottom row).

integral; for instance,

$$E_x(t) = \int_{-\infty}^{\infty} A_x(\nu)\ \mathrm{e}^{-2\pi \mathrm{i}\nu t} \mathrm{d}\nu. \tag{3.1}$$

With our previous choice of complex exponentials with a minus sign in the argument for denoting harmonic functions, Eq. (3.1) can be interpreted to mean the x component of the electric field being the inverse Fourier transform of $A_x(\nu)$; in other words, $A_x(\nu)$ is the spectrum or Fourier transform of the x component of the electric field. According to Fourier theory, $E_x(t)$ and $A_x(\nu)$ provide the same information: they are different representations of the very same function; the function in the time domain tells us about variations with time of the electric field, while the spectrum tells us the frequency content of the field. As illustrated in Fig. 3.1, when passing from a monochromatic beam of frequency ν_0 to a polychromatic beam of central frequency ν_0, the spectrum of light changes from a Dirac delta distribution to a distribution with non-zero values in an interval of width $\Delta\nu$, which is usually much smaller than ν_0. Although Fig. 3.1 represents only the amplitude spectrum, the phase spectrum should also be considered because $A_x(\nu)$ is generally complex. A similar finite-bandwidth spectrum can be found if one observes an (ideal) monochromatic wave during a (necessarily) finite time window. The convolution theorem ensures that the observed spectrum is the convolution of the spectrum of the monochromatic wave (a Dirac delta) and that of the window (a sinc function if the window is a top-hat function).

The typical time-scales of such a polychromatic electromagnetic wave are the *mean period*, $\tau_0 = 1/\nu_0$, and the *coherence time*, $\tau_\mathrm{c} = 1/\Delta\nu$. In practice, these scales are much smaller than the (finite) time interval for measurements, τ_m. For example, a polychromatic beam of width 10 nm around 500 nm has a mean period

$\tau_0 = 1.6 \times 10^{-15}$ s and a coherence time $\tau_c = 8.3 \times 10^{-14}$ s. It is then clear that even the fastest electronic detectors with τ_m in the order of micro- or nanoseconds are unable to follow the rapidly varying electric field fluctuations. Measurements, then, are averages of the type

$$\frac{1}{\tau_m} \int_{-\frac{\tau_m}{2}}^{\frac{\tau_m}{2}} E_x(t) E_x^*(t) \mathrm{d}t. \tag{3.2}$$

Since $\tau_m \gg \tau_0$ and τ_c, it is convenient to let τ_m approach infinity and then representing measurements by

$$\langle E_x E_x^* \rangle \equiv \lim_{\tau_m \to \infty} \frac{1}{\tau_m} \int_{-\infty}^{\infty} E_{x,\tau_m}(t) E_{x,\tau_m}^*(t) \mathrm{d}t, \tag{3.3}$$

where, by definition,

$$E_{x,\tau_m}(t) \equiv E_x(t)\, \Pi_{\tau_m}(t), \tag{3.4}$$

with

$$\Pi_{\tau_m}(t) = \begin{cases} 1 & -(\tau_m/2) \le t \le (\tau_m/2), \\ 0 & \text{otherwise.} \end{cases} \tag{3.5}$$

After assuming the local square integrability of $E_x(t)$ and light as a statistically ergodic process (see, for example, Born and Wolf, 1993), it can be shown that a finite, non-zero, limit exists for the right-hand side of Eq. (3.3). Similar limits can also be found for $\langle E_y E_y^* \rangle$, $\langle E_x E_y^* \rangle$, and $\langle E_y E_x^* \rangle$.

3.2 The quasi-monochromatic plane wave

The x (and y) component of the electric field of a polychromatic beam streaming along the Z axis can always be put in the form

$$E_x(t) = \left[\mathcal{E}_x(t)\ \mathrm{e}^{\mathrm{i}\phi_x(t)}\right] \mathrm{e}^{-2\pi \mathrm{i} \nu_0 t}, \tag{3.6}$$

that is, as a monochromatic plane wave of frequency equal to its mean frequency modulated by a (complex) amplitude that varies with time. In Eq. (3.6), obviously,

$$\begin{aligned} \mathcal{E}_x(t) &= \sqrt{E_x(t) E_x^*(t)} = |E_x(t)|, \\ \phi_x(t) &= 2\pi \nu_0 t + \tan^{-1} \left\{ \frac{\mathrm{Im}\,[E_x(t)]}{\mathrm{Re}\,[E_x(t)]} \right\}. \end{aligned} \tag{3.7}$$

Equations (3.1) and (3.6) imply that

$$\mathcal{E}_x(t)\, \mathrm{e}^{-\mathrm{i}[2\pi \nu_0 t - \phi_x(t)]} = \int_{-\infty}^{\infty} A_x(\nu)\, \mathrm{e}^{-2\pi \mathrm{i} \nu t} \mathrm{d}\nu, \tag{3.8}$$

and, using the variable change $\mu \equiv \nu - \nu_0$,

$$\mathcal{E}_x(t)\,\mathrm{e}^{\mathrm{i}\phi_x(t)} = \int_{-\infty}^{\infty} A_x(\mu + \nu_0)\,\mathrm{e}^{-2\pi\mathrm{i}\mu t}\,\mathrm{d}\mu, \tag{3.9}$$

which gives us the spectral content of the time-dependent amplitude and phase of the plane wave. Since $A_x(\mu + \nu_0)$ is the Fourier transform of the function on the left-hand side, the wider A_x is, the faster the variation of the modulating envelope.

If $\Delta\nu/\nu_0 \ll 1$ then $A_x(\mu + \nu_0)$ differs from zero only in the neighborhood of $\mu = 0$. In other words, the amplitude and phase of the polychromatic wave have only low frequencies. Hence, the left-hand side of Eq. (3.9) must be interpreted as a *slowly varying* function of t. In such a case, the polychromatic plane wave is said to be *quasi-monochromatic*.

3.3 The polarization tensor and the Stokes parameters of a quasi-monochromatic plane wave

From a formal point of view, Eq. (3.6) allows us to consider a quasi-monochromatic plane wave as a purely monochromatic plane wave with the only difference that both amplitude and phase are (slowly) time-dependent.

The energy quantities of the monochromatic wave, enclosed in the polarization tensor, now become time averages like those of Eq. (3.3). Hence, one can define the coherency matrix of a polychromatic (in particular, quasi-monochromatic) light beam as

$$\mathbf{C} \equiv \begin{pmatrix} \langle E_x(t)E_x^*(t)\rangle & \langle E_x(t)E_y^*(t)\rangle \\ \langle E_y(t)E_x^*(t)\rangle & \langle E_y(t)E_y^*(t)\rangle \end{pmatrix}, \tag{3.10}$$

and, using Eq. (3.6),

$$\mathbf{C} = \begin{pmatrix} \langle \mathcal{E}_x^2(t)\rangle & \langle \mathcal{E}_x(t)\mathcal{E}_y(t)\;\mathrm{e}^{\mathrm{i}\phi(t)}\rangle \\ \langle \mathcal{E}_x(t)\mathcal{E}_y(t)\;\mathrm{e}^{-\mathrm{i}\phi(t)}\rangle & \langle \mathcal{E}_y^2(t)\rangle \end{pmatrix}, \tag{3.11}$$

where $\phi(t) \equiv \phi_x(t) - \phi_y(t)$.

As in the monochromatic case, the elements of the coherency matrix (and hence the energy content of the electromagnetic field) are independent of both time (after the assumption of stationarity) and space in the absence of sources and sinks. This invariance is equivalent to that noted in classical textbooks on stellar atmospheres and radiative transfer in astrophysics for the specific intensity. The trace of the polarization tensor is still the total intensity of the light beam:

$$\mathrm{Tr}(\mathbf{C}) = \left\langle \mathcal{E}_x^2(t)\right\rangle + \left\langle \mathcal{E}_y^2(t)\right\rangle \geq 0. \tag{3.12}$$

However, the determinant of the polarization tensor is not necessarily zero. In fact, Schwarz's inequality requires that

$$|C_{ij}| \leq \sqrt{C_{ii}C_{jj}}, \quad i \neq j, \tag{3.13}$$

or, in other terms,

$$\det(\mathbf{C}) = C_{11}C_{22} - C_{12}C_{21} \geq 0. \tag{3.14}$$

The non-diagonal terms of the coherency matrix are generally complex, but they are conjugates of each other (the polarization tensor is a Hermitian matrix). They provide a measure of the correlation between the x and y components of the electric field vector:† the smaller the determinant, the closer the beam is to being coherent (i.e., monochromatic) radiation.

The Stokes parameters of the quasi-monochromatic plane wave turn out to be:

$$\begin{aligned} I &= \kappa\left(\langle \mathcal{E}_x^2\rangle + \langle \mathcal{E}_y^2\rangle\right), \\ Q &= \kappa\left(\langle \mathcal{E}_x^2\rangle - \langle \mathcal{E}_y^2\rangle\right), \\ U &= 2\kappa\langle \mathcal{E}_x\mathcal{E}_y\cos\phi(t)\rangle, \\ V &= 2\kappa\langle \mathcal{E}_x\mathcal{E}_y\sin\phi(t)\rangle, \end{aligned} \tag{3.15}$$

where we have used the same combinations of the coherency matrix elements as for Eqs (2.20). Note that all four Stokes parameters are real and have dimensions of energy. They are thus *measurable*. The actual value of κ is irrelevant in practice since only Q/I, U/I, and V/I are sought in most cases.

Equation (3.15) can be similarly derived from a more appealingly symmetric definition of the Stokes parameters:

$$\begin{aligned} I &= \kappa\,\mathrm{Tr}\,(\mathbf{C}\boldsymbol{\sigma}_0), \\ Q &= \kappa\,\mathrm{Tr}\,(\mathbf{C}\boldsymbol{\sigma}_1), \\ U &= \kappa\,\mathrm{Tr}\,(\mathbf{C}\boldsymbol{\sigma}_2), \\ V &= \kappa\,\mathrm{Tr}\,(\mathbf{C}\boldsymbol{\sigma}_3), \end{aligned} \tag{3.16}$$

† The correlation between the square integrable functions $f(t)$ and $g(t)$ is defined as $\int_{-\infty}^{\infty} f(t+t')g^*(t')\mathrm{d}t'$ so that it is in fact a function of the displacement, t. The coherency matrix elements are then correlations for zero displacement.

where σ_i $(i = 0, 1, 2, 3)$ are the 2×2 identity matrix and the three Pauli matrices:‡

$$\begin{aligned} \boldsymbol{\sigma}_0 &= \begin{pmatrix} 1 & 0 \\ 0 & 1 \end{pmatrix}, & \boldsymbol{\sigma}_1 &= \begin{pmatrix} 1 & 0 \\ 0 & -1 \end{pmatrix}, \\ \boldsymbol{\sigma}_2 &= \begin{pmatrix} 0 & 1 \\ 1 & 0 \end{pmatrix}, & \boldsymbol{\sigma}_3 &= \begin{pmatrix} 0 & \mathrm{i} \\ -\mathrm{i} & 0 \end{pmatrix}. \end{aligned} \tag{3.17}$$

The consistency of Eq. (3.16) can easily be checked from Eq. (2.21).

Therefore, I, Q, U, and V are indeed the coefficients of an expansion of the coherency matrix in terms of the $\boldsymbol{\sigma}_i$ matrices:

$$\mathbf{C} = \frac{1}{2\kappa}(I\boldsymbol{\sigma}_0 + Q\boldsymbol{\sigma}_1 + U\boldsymbol{\sigma}_2 + V\boldsymbol{\sigma}_3). \tag{3.18}$$

3.4 Degree of polarization and the Poincaré sphere

From the binding condition (3.14) on $\mathbf{C}$ and the definition of the Stokes parameters (3.15) or (3.16), it is easy to see that

$$I^2 - Q^2 - U^2 - V^2 \geq 0, \tag{3.19}$$

and from the definite non-negativeness of $\mathrm{Tr}(\mathbf{C})$ [Eq. (3.12)],

$$I \geq 0. \tag{3.20}$$

Conditions (3.19) and (3.20) must be verified by any set of physically meaningful Stokes parameters; that is, by I, Q, U, and V values of any quasi-monochromatic plane wave. The interpretation of inequality (3.20) is simple and is indeed the same as for the monochromatic case: negative intensities do not exist; the particular case when $I = 0$ corresponds to the absence of light. Condition (3.19), however, differs from Eq. (2.22) for monochromatic light: a quasi-monochromatic plane wave may or may not be totally polarized. From Eq. (3.11), it is easy to see that equality between the square of I and the sum of the squares of Q, U, and V holds if, and only if, $\mathcal{E}_y(t)/\mathcal{E}_x(t)$ and $\phi(t)$ are constant. In such a case, the quasi-monochromatic light is said to be *totally polarized* and *cannot be distinguished from monochromatic light as far as polarization is concerned.* When $Q = U = V = 0$, the light is said to be *natural* or *completely unpolarized.* Otherwise, the light is said to be *partially polarized.* We shall come to understand the meaning of this term later in this section.

‡ The "standard" Pauli matrices are $\sigma_{s,1} = \sigma_2$, $\sigma_{s,2} = -\sigma_3$, and $\sigma_{s,3} = \sigma_1$. Again, sign conventions and the (also arbitrary) ordering of the Stokes parameters make us choose the matrices of Eq. (3.17).

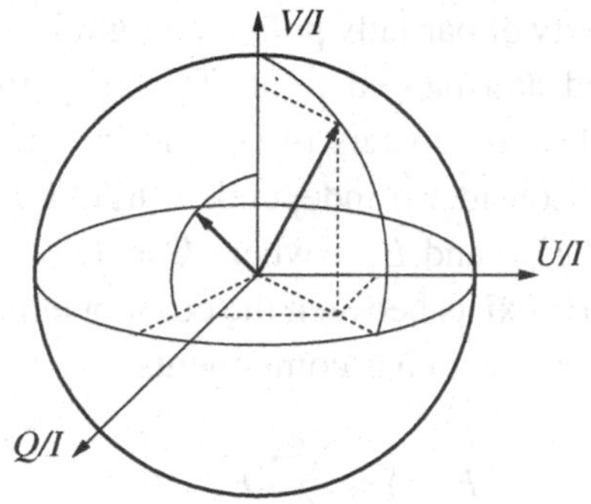

Fig. 3.2. Quasi-monochromatic light is partially polarized in general. Its polarization vector lies within the Poincaré sphere. The center of the sphere corresponds to natural light. The surface of the sphere corresponds to totally polarized light.

Given a set of four Stokes parameters, one can define a *polarization vector*,

$$\boldsymbol{p} \equiv \left(\frac{Q}{I}, \frac{U}{I}, \frac{V}{I}\right)^{\mathrm{T}}, \tag{3.21}$$

where the index "T" denotes the transposition operation. Thus, every beam of light can be characterized by its intensity and its polarization vector.

According to inequality (3.19), $\boldsymbol{p}$ belongs to the Poincaré sphere (see Fig. 3.2) since

$$0 \leq p = \frac{\sqrt{Q^2 + U^2 + V^2}}{I} \leq 1 \Longrightarrow \boldsymbol{p} \in \mathbb{P}, \tag{3.22}$$

disregarding the case of $I = 0$. The magnitude, p, of the polarization vector is called the *degree of polarization* of the quasi-monochromatic beam.

If $p = 0$, then $Q = U = V = 0$. This case corresponds to natural light; it is geometrically located at the center of the Poincaré sphere. The physical reasons for having Q, U, and V equal to zero are twofold. First, natural light has the same intensity in every direction perpendicular to the direction of propagation. Second, it is not altered by any previous retardation (phase addition) of any of the Cartesian components of the electric field vector. Thus, unpolarized light presents no preferential motion of $\boldsymbol{E}_{\mathrm{R}}$ on a plane perpendicular to the direction of propagation; in other words, all possible motions of $\boldsymbol{E}_{\mathrm{R}}$ have the same probability of occurring. If $p = 1$, then $I^2 = Q^2 + U^2 + V^2$. This case corresponds to totally polarized light. Its polarization vector ends on the surface of the Poincaré sphere and its electric vector can be thought of as undergoing an elliptic motion in exactly the same way as purely monochromatic light. The internal points in the sphere represent all possible states of partially polarized light.

A very important property of partially polarized light is that it can be decomposed into completely polarized and natural light. To verify this property, let us first demonstrate that the Stokes parameters of several independent (incoherent) light beams can be summed.† Consider n independent light beams whose electric field vectors have components $E_{x,k}$ and $E_{y,k}$, where $k = 1, 2, \ldots, n$. By "independent" we mean that no correlation exists between the components of different beams. The electric field of the resulting beam has components

$$E_x(t) = \sum_{k=1}^{n} E_{x,k}(t) \tag{3.23}$$

and

$$E_y(t) = \sum_{k=1}^{n} E_{y,k}(t). \tag{3.24}$$

The coherency matrix of the ensemble field is then given by

$$\begin{aligned} C_{ij} &= \langle E_i(t) E_j^*(t) \rangle = \sum_{k=1}^{n} \sum_{l=1}^{n} \langle E_{i,k}(t) E_{j,l}^*(t) \rangle \\ &= \sum_{k=1}^{n} \langle E_{i,k}(t) E_{j,k}^*(t) \rangle + \sum_{k \neq l} \langle E_{i,k}(t) E_{j,l}^*(t) \rangle. \end{aligned} \tag{3.25}$$

Since the last term of Eq. (3.25) must be zero by hypothesis (no correlation may exist between incoherent light beams),‡ it turns out that

$$C_{ij} = \sum_{k=1}^{n} \langle E_{i,k}(t) E_{j,k}^*(t) \rangle = \sum_{k=1}^{n} C_{ij,k}; \tag{3.26}$$

i.e., the coherency matrix elements of the sum of independent beams is the sum of the coherency matrix elements of the individual beams. Since the Stokes parameters are linear combinations of C_{ij}, this property readily holds for I, Q, U, and V.

It is important to take note that Eq. (3.26) is valid only for the incoherent superposition of polarized light. The Stokes formalism is unable to deal with coherent superposition of completely polarized light because of the quadratic nature of I, Q, U, and V. Coherent superposition of totally polarized light such as occurs in interference or diffraction phenomena, requires the addition of amplitudes of the electric field. Such phenomena are studied within the Jones formalism, which in turn is unable to describe partially polarized states. Virtually all the problems discussed in this book are more suitably described within the Stokes formalism, and

† This and the remaining results of this section were found as early as 1852 by G. G. Stokes.

‡ Remember the definition of correlation in the first footnote in this chapter (page 23).

Table 3.1. *Stokes vectors for some completely polarized states. Angles are referred to the positive X axis counter-clockwise*

Polarization state	Stokes vector
Natural	$(1, 0, 0, 0)^{\mathrm{T}}$
Linear at 0°	$(1, 1, 0, 0)^{\mathrm{T}}$
Linear at 90°	$(1, -1, 0, 0)^{\mathrm{T}}$
Linear at 45°	$(1, 0, 1, 0)^{\mathrm{T}}$
Linear at 135°	$(1, 0, -1, 0)^{\mathrm{T}}$
Right-handed circular	$(1, 0, 0, 1)^{\mathrm{T}}$
Left-handed circular	$(1, 0, 0, -1)^{\mathrm{T}}$

the Jones formalism is therefore ignored here. The interested reader is referred to Shurcliff (1962) for the original bibliography on the Jones calculus.

It is customary (and useful) to group all four Stokes parameters in a four-vector

$$\boldsymbol{I} \equiv (I, Q, U, V)^{\mathrm{T}}, \tag{3.27}$$

the so-called *Stokes vector*. Examples of such Stokes vectors for particular cases of totally polarized states can be found in Table 3.1, where intensities have been normalized to unity. The meaning of the labels in the first column will be made clear later (Section 3.5).

The property of addition for the Stokes parameters of independent light beams is given by

$$\boldsymbol{I} = \sum_{k=1}^{n} \boldsymbol{I}_k, \tag{3.28}$$

where $\boldsymbol{I}_k$ corresponds to the Stokes vector of the kth independent component. Using this property, one easily gets

$$\begin{pmatrix} I \\ Q \\ U \\ V \end{pmatrix} = \begin{pmatrix} \sqrt{Q^2 + U^2 + V^2} \\ Q \\ U \\ V \end{pmatrix} + \begin{pmatrix} I - \sqrt{Q^2 + U^2 + V^2} \\ 0 \\ 0 \\ 0 \end{pmatrix}. \tag{3.29}$$

Therefore, any quasi-monochromatic light beam can be considered as the (incoherent) sum of a completely polarized beam [the first term on the right-hand side of Eq. (3.29)] and a completely unpolarized beam [the second term on the right-hand side of Eq. (3.29)]:

$$\boldsymbol{I} = \boldsymbol{I}_{\mathrm{pol}} + \boldsymbol{I}_{\mathrm{nat}}. \tag{3.30}$$

Hence, the quasi-monochromatic light is always partially polarized. The two extreme cases of totally polarized or unpolarized light correspond to those for which one of the two terms of Eq. (3.29) is zero. Equation (3.30) includes

$$I = I_{\text{pol}} + I_{\text{nat}}, \tag{3.31}$$

which implies that

$$p = \frac{I_{\text{pol}}}{I}; \tag{3.32}$$

that is, the degree of polarization is the ratio between the intensity of the completely polarized component and the total intensity of the beam. This is the reason for the name of the magnitude of the polarization vector, p.

If intensity is thought of as a scaling factor, the polarization state of any light beam is completely described by the polarization vector, $\boldsymbol{p}$. Reminiscent of this geometrical representation, some polarization terms are borrowed from geometry. Among these, it is convenient to introduce here the concept of orthogonally polarized states. Contrary to what one might expect, two orthogonal states have their polarization vectors not orthogonal but anti-parallel: $\boldsymbol{p}$ and $-\boldsymbol{p}$ are orthogonal states. The reason for this paradox is not unexpected, however. The qualifier "orthogonal" refers to the way polarization is measured. In particular, any two orthogonal states are detected by a device whose characteristics differ in just one direction, which, for one of the states, is perpendicular to that for the other state (see Section 3.5 for details).

The addition property (3.28) allows us to consider natural light as the (incoherent) sum of any two orthogonally polarized states, each having half the intensity of the original beam:

$$\begin{pmatrix} I \\ 0 \\ 0 \\ 0 \end{pmatrix} = \frac{1}{2}\begin{pmatrix} I \\ Q \\ U \\ V \end{pmatrix} + \frac{1}{2}\begin{pmatrix} I \\ -Q \\ -U \\ -V \end{pmatrix}. \tag{3.33}$$

For instance, natural light can be thought of either as being the sum of two linearly polarized lights at 0° and 90°,

$$\begin{pmatrix} I \\ 0 \\ 0 \\ 0 \end{pmatrix} = \frac{1}{2}\begin{pmatrix} I \\ I \\ 0 \\ 0 \end{pmatrix} + \frac{1}{2}\begin{pmatrix} I \\ -I \\ 0 \\ 0 \end{pmatrix}, \tag{3.34}$$

or at 45° and 135°,

$$\begin{pmatrix} I \\ 0 \\ 0 \\ 0 \end{pmatrix} = \frac{1}{2}\begin{pmatrix} I \\ 0 \\ I \\ 0 \end{pmatrix} + \frac{1}{2}\begin{pmatrix} I \\ 0 \\ -I \\ 0 \end{pmatrix}, \tag{3.35}$$

or of two circularly (right-handed and left-handed) polarized components,

$$\begin{pmatrix} I \\ 0 \\ 0 \\ 0 \end{pmatrix} = \frac{1}{2}\begin{pmatrix} I \\ 0 \\ 0 \\ I \end{pmatrix} + \frac{1}{2}\begin{pmatrix} I \\ 0 \\ 0 \\ -I \end{pmatrix}. \tag{3.36}$$

The decomposition of both partially polarized [Eq. (3.29)] and natural light [Eq. (3.33)] thus enables us in the end to consider every polarization state as the sum of totally polarized states and is very helpful in understanding the measurement procedures of the Stokes parameters.

Another interesting consequence of the addition property is that the incoherent sum of a totally polarized beam plus a partially polarized beam is always partially polarized. Let $\boldsymbol{I}_1$ be the Stokes vector of the totally polarized beam, such that $I_1^2 = Q_1^2 + U_1^2 + V_1^2$, and $\boldsymbol{I}_2$ be the Stokes vector of the partially polarized beam, such that $I_2^2 > Q_2^2 + U_2^2 + V_2^2$. The resulting beam has a Stokes vector, $\boldsymbol{I}_3$, such that

$$Q_3^2 + U_3^2 + V_3^2 = (Q_1 + Q_2)^2 + (U_1 + U_2)^2 + (V_1 + V_2)^2$$

and, by the triangular inequality, the right-hand side of the above equation is less than $\left(\sqrt{Q_1^2 + U_1^2 + V_1^2} + \sqrt{Q_2^2 + U_2^2 + V_2^2}\right)^2$. Therefore, we finally obtain

$$Q_3^2 + U_3^2 + V_3^2 < (I_1 + I_2)^2 = I_3^2. \tag{3.37}$$

3.5 Measuring the polarization state of quasi-monochromatic light

To gain an insight into the physical meaning of the Stokes parameters it is necessary to grasp how they can be measured. If polarization means a definite motion of the electric field vector, one should account for directions of motion and phase differences between the two Cartesian components of the field. This is accomplished by means of two specific measuring devices: the linear analyzer and the linear retarder.

An optical system is said to be a *linear analyzer* if it presents maximum transmission for the component E_θ of the electric field in a direction forming an angle θ with the positive X axis and completely absorbs (or reflects) the component $E_{\theta+\pi/2}$

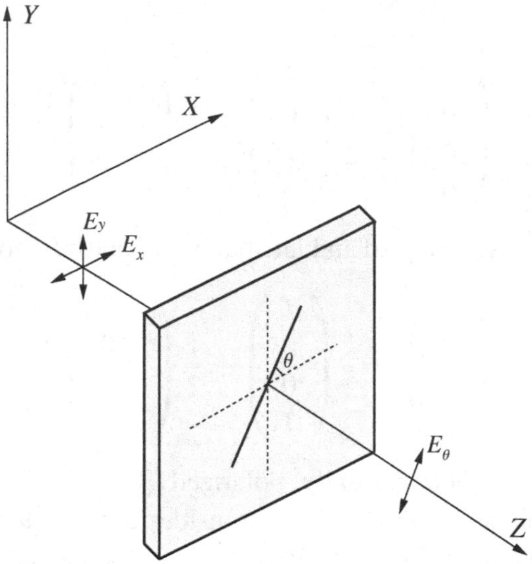

Fig. 3.3. Sketch of linear analyzer–polarizer behavior. Input light has two Cartesian components in general. Only the projections of these two components over the optical axis (at an angle θ to the X axis) are transmitted. The resulting beam from the linear analyzer–polarizer is completely linearly polarized along the direction of the optical axis.

of the electric field in the orthogonal direction (see Fig. 3.3). The direction of angle θ is called the *optical axis* of the linear analyzer. In other words, a linear analyzer completely transmits any light beam linearly polarized along its optical axis and completely extinguishes any light beam perpendicularly linearly polarized to its optical axis. Most texts use the term "linear polarizer" for describing this device. There is no possible confusion because both systems are the same physical device. The two terms conform to the different roles such a device may play in polarization optics: a linear polarizer is an optical system such that, after interaction with it, light becomes completely linearly polarized. At the exit of such a device, the new x and y components of the electric field corresponding to an arbitrary input are

$$E'_x = E_x \cos\theta; \quad E'_y = E_y \sin\theta. \tag{3.38}$$

An optical system is said to be a *linear retarder* if it imparts a *retardance* (a phase lag) δ to one of the orthogonal components of $\boldsymbol{E}$ with respect to the other. The electric field of the retarded component is colinear with the so-called *slow axis* and the other is parallel to the *fast axis* of the retarder (see Fig. 3.4). If X is the fast axis, the x and y components of the outgoing electric field are

$$E'_x = E_x; \quad E'_y = E_y\, \mathrm{e}^{\mathrm{i}\delta}. \tag{3.39}$$

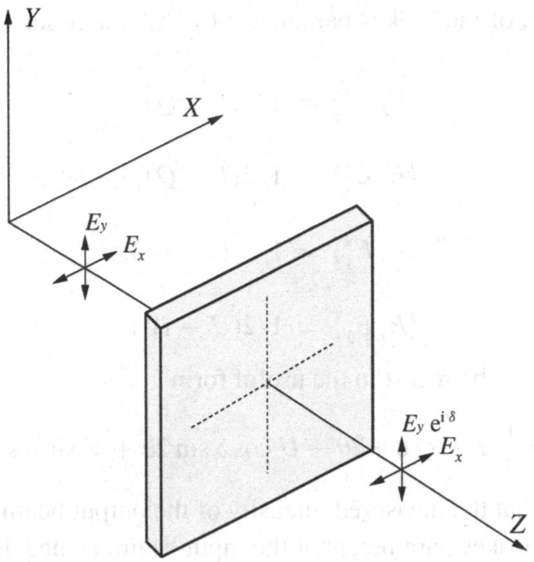

Fig. 3.4. A linear retarder shifts the phase of that Cartesian component of the input which is linearly polarized along its slow axis. In this example, X is the fast axis.

Imagine a quasi-monochromatic plane wave to be transmitted through a linear retarder and then through a linear analyzer like those described above. Let us study the (measurable) intensity of the transmitted light beam that will obviously depend on both θ and δ.

According to Eqs (3.39) and (3.38), at the exit of the analyzer the light is completely linearly polarized at an angle θ, the amplitude of the electric vector being given by

$$E_\theta(t;\delta) = E_x \cos\theta + E_y \sin\theta \, \mathrm{e}^{\mathrm{i}\delta}. \tag{3.40}$$

The intensity of the output beam is

$$I_{\mathrm{meas}}(\theta,\delta) = \langle E_\theta(t;\delta) E_\theta^*(t;\delta)\rangle, \tag{3.41}$$

or, written in full (but excluding functional dependences),

$$\begin{aligned} I_{\mathrm{meas}}(\theta,\delta) = \langle & E_x E_x^* \cos^2\theta + E_y E_y^* \sin^2\theta \\ & + \frac{1}{2} E_x E_y^* \sin 2\theta \, \mathrm{e}^{-\mathrm{i}\delta} + \frac{1}{2} E_x^* E_y \sin 2\theta \, \mathrm{e}^{\mathrm{i}\delta}\rangle. \end{aligned} \tag{3.42}$$

From the definition of the Stokes parameters (3.15), and assuming that $\kappa = 1$, we easily obtain:

$$\begin{aligned}\langle E_x E_x^* \rangle &= 1/2(I + Q),\\ \langle E_y E_y^* \rangle &= 1/2(I - Q),\\ \langle E_x E_y^* \rangle &= 1/2(U + \mathrm{i}V),\\ \langle E_x^* E_y \rangle &= 1/2(U - \mathrm{i}V).\end{aligned} \tag{3.43}$$

Hence, Eq. (3.42) can be recast in the useful form

$$I_{\mathrm{meas}}(\theta, \delta) = \frac{1}{2}(I + Q\cos 2\theta + U\cos\delta\sin 2\theta + V\sin\delta\sin 2\theta). \tag{3.44}$$

We have found that the measured intensity of the output beam is a linear combination of the four Stokes parameters of the input beam. Hence, by varying θ and δ one can easily determine I, Q, U, and V. Specifically,

$$\begin{aligned}I &= I_{\mathrm{meas}}(0, 0) + I_{\mathrm{meas}}(\pi/2, 0),\\ Q &= I_{\mathrm{meas}}(0, 0) - I_{\mathrm{meas}}(\pi/2, 0),\\ U &= I_{\mathrm{meas}}(\pi/4, 0) - I_{\mathrm{meas}}(3\pi/4, 0),\\ V &= I_{\mathrm{meas}}(\pi/4, \pi/2) - I_{\mathrm{meas}}(3\pi/4, \pi/2).\end{aligned} \tag{3.45}$$

Note that Stokes Q, U, and V result from differences in two intensity measurements for which the optical axis of the analyzer has been rotated by $\pi/2$. For example, that light beam for which only $I_{\mathrm{meas}}(0, 0)$ is different from zero is in a polarization state orthogonal to the other beam, for which only $I_{\mathrm{meas}}(\pi/2, 0)$ is different from zero. As commented on in the previous section, the two orthogonal states have anti-parallel polarization vectors.

Let us digress a little on the physical meaning of Eqs (3.45). Let us consider the most general input, a partially polarized beam, and explore what is going on after the measurements described in Eqs (3.45). According to Eq. (3.30), the input beam is always the sum of a natural beam plus a totally polarized beam. Now, the natural component can be decomposed into two orthogonally polarized beams [Eq. (3.33)], one of which will be completely transmitted and the other completely absorbed or reflected. Fifty percent of the intensity of the natural component, then, is contributed to every measurement. Since the equations for Q, U, and V are differences, such a natural component cancels out for all three parameters and contributes only to Stokes I – the total intensity of the input beam. Hence, polarization information is provided by Q, U, and V. As will be shown in Section 4.6.5, the compound device (linear retarder plus linear analyzer) is indeed an analyzer which completely transmits one state of polarization (not necessarily linear) and completely extinguishes the orthogonal state.† According to this interpretation, if we denote the state of the analyzer by (θ, δ), we can identify $(0, 0)$ as the analyzer that completely transmits the linearly polarized component along the X axis; $(\pi/2, 0)$ as the analyzer that completely transmits the linearly polarized component along the Y axis; $(\pi/4, 0)$ as the analyzer that completely transmits the linearly polarized component in the 45° direction; $(3\pi/4, 0)$ as the analyzer that completely transmits the linearly polarized component in the 135° direction; $(\pi/4, \pi/2)$ as the analyzer that completely transmits the right-handed circularly polarized component; and, finally, $(3\pi/4, \pi/2)$ as the analyzer that completely transmits the left-handed circularly polarized component of the incoming beam of radiation. Thus, the interpretation of the Stokes parameters follows naturally: I is the total intensity; Q is the difference between the intensities of linear components at 0° and 90°; U is the difference between the intensities of linear components at 45° and 135°; and V is the difference between the intensities of the right-handed and left-handed circularly polarized components of the incoming quasi-monochromatic plane wave (see Table 3.1).

3.6 A further perspective on polarization properties

So far we have been dealing with a Cartesian description of the electromagnetic field that has led to a specific definition of the Stokes parameters but other descriptions are possible. Within the preceding framework, the interpretation of Stokes Q as the difference between the intensities of two linearly polarized components along the X and Y axes followed naturally from definition (3.15) itself. Likewise, a counter-clockwise rotation of 45° of the axes would have led to a similar interpretation for Stokes U. The interpretation of Stokes V, however, needed the considerations concerning the measuring devices of the preceding section to be made clear: the fourth Stokes parameter is the intensity difference of two circularly polarized components (right-handed minus left-handed).

There is another representation of the electric field of quasi-monochromatic light in which the physical interpretation of Stokes V is as natural as that for Stokes Q and U in the Cartesian representation. This alternative view is more suitable for a number of problems. In particular, it is very useful in the analysis of radiative transfer through anisotropic media and is thus very relevant for the second part of the book (Chapter 7 onwards), so we now describe it.

† Note that measuring I, Q, and U needs no retarder ($\delta = 0$) and the compound analyzer is thus linear.

Instead of representing the three-dimensional space by an orthonormal set of Cartesian vectors $\{\hat{\boldsymbol{x}}, \hat{\boldsymbol{y}}, \hat{\boldsymbol{z}}\}$, let us use the new set of *complex* vectors

$$\begin{aligned} \hat{\boldsymbol{l}} &\equiv \tfrac{1}{\sqrt{2}}(\hat{\boldsymbol{x}} + \mathrm{i}\hat{\boldsymbol{y}}), \\ \hat{\boldsymbol{r}} &\equiv \tfrac{1}{\sqrt{2}}(\hat{\boldsymbol{x}} - \mathrm{i}\hat{\boldsymbol{y}}), \\ \hat{\boldsymbol{z}} &\equiv \hat{\boldsymbol{z}}. \end{aligned} \tag{3.46}$$

This new set is also orthonormal for the scalar product $\boldsymbol{a} \cdot \boldsymbol{b}^*$, where $\boldsymbol{a}$ and $\boldsymbol{b}$ are any three-dimensional vectors.

The reader may have already guessed that the parameters l and r refer to left-handed and right-handed circularly polarized components respectively. As a matter of fact,

$$\hat{\boldsymbol{l}} = \frac{1}{\sqrt{2}}(\hat{\boldsymbol{x}} + \mathrm{e}^{\mathrm{i}\pi/2}\hat{\boldsymbol{y}}), \tag{3.47}$$

so that $\hat{\boldsymbol{l}}$ represents an electric field whose x and y components have equal amplitude (and thus intensity) and for which the difference in phase between the x and y components is $-\pi/2$. Hence, $\hat{\boldsymbol{l}}$ is the unit vector for left-handed circularly polarized radiation, according to our convention for handedness. In the same way,

$$\hat{\boldsymbol{r}} = \frac{1}{\sqrt{2}}(\hat{\boldsymbol{x}} + \mathrm{e}^{-\mathrm{i}\pi/2}\hat{\boldsymbol{y}}), \tag{3.48}$$

so that $\hat{\boldsymbol{r}}$ represents an electric field whose x and y components have equal amplitude (and thus intensity) and for which the difference in phase between the x and y components is $\pi/2$. Therefore, $\hat{\boldsymbol{r}}$ is the unit vector for right-handed circularly polarized radiation.

In the basis of right-handed and left-handed circularly polarized vectors, the electric field of a general quasi-monochromatic plane wave can be written as

$$\boldsymbol{E}(t) = E_r(t)\hat{\boldsymbol{r}} + E_l(t)\hat{\boldsymbol{l}}, \tag{3.49}$$

where we still assume that $\hat{\boldsymbol{z}}$ represents the wavefront normal. Equation (3.49) does indeed decompose the electric field into two orthogonally polarized waves, much in the same way as in Eq. (3.36). The new components can easily be identified in Cartesian terms. They verify that

$$\begin{aligned} E_x(t) &= \tfrac{1}{\sqrt{2}}\left[E_r(t) + E_l(t)\right], \\ E_y(t) &= \tfrac{\mathrm{i}}{\sqrt{2}}\left[E_l(t) - E_r(t)\right]. \end{aligned} \tag{3.50}$$

Equations (3.50) allow us to rewrite the coherency matrix elements as

$$
\begin{aligned}
C_{11} &= \tfrac{1}{2}\langle |E_r|^2 + |E_l|^2 + E_l E_r^* + E_r E_l^* \rangle, \\
C_{12} &= \tfrac{\mathrm{i}}{2}\langle |E_r|^2 - |E_l|^2 + E_l E_r^* - E_r E_l^* \rangle, \\
C_{21} &= \tfrac{\mathrm{i}}{2}\langle |E_l|^2 - |E_r|^2 + E_l E_r^* - E_r E_l^* \rangle, \\
C_{22} &= \tfrac{1}{2}\langle |E_r|^2 + |E_l|^2 - E_l E_r^* - E_r E_l^* \rangle,
\end{aligned}
\tag{3.51}
$$

where the explicit dependence of E_r and of E_l on t has been omitted for simplicity. With these expressions for **C**, the Stokes parameters are

$$
\begin{aligned}
I &= \kappa\langle |E_r|^2 + |E_l|^2 \rangle, \\
Q &= \kappa(\langle E_r E_l^* \rangle + \langle E_l E_r^* \rangle), \\
U &= \mathrm{i}\kappa(\langle E_l E_r^* \rangle - \langle E_r E_l^* \rangle), \\
V &= \kappa\langle |E_r|^2 - |E_l|^2 \rangle.
\end{aligned}
\tag{3.52}
$$

Therefore, the fourth Stokes parameter is naturally seen as the difference between the intensities of right-handed and left-handed polarized components of the quasi-monochromatic light beam.

Recommended bibliography

Born, M. and Wolf, E. (1993). *Principles of optics*, 6th edition (Pergamon Press: Oxford). Chapter 10. Sections 10.2–10.3, 10.7.3, and 10.8.

Chandrasekhar, S. (1947). On the radiative equilibrium of a stellar atmosphere. XV. *Astrophys. J.* **105**, 424.

Chandrasekhar, S. (1960). *Radiative transfer* (Dover Publications, Inc.: New York). Chapter 1. Sections 15.2–15.4.

Shurcliff, W.A. (1962). *Polarized light* (Harvard University Press & Oxford University Press: Cambridge, Mass. & London). Bibliography.

Stokes, G.G. (1852). On the composition and resolution of streams of polarized light from different sources. *Trans. Cambridge Phil. Soc.* **9**, 399.

4

Linear optical systems acting on polarized light

Porque aquellas cosas que bien no son pensadas, aunque algunas veces hayan buen fin, comúnmente crían desvariados efectos. Así que la mucha especulación nunca carece de buen fruto.

—*Fernando de Rojas, 1514.*

For those matters that are ill thought out may yet end well, even though they often breed strange consequences. Hence, much speculation never fails to bring forth some good fruit.

This chapter is aimed at understanding how nature and laboratory devices may change the polarization state of light. The transformations of the Stokes parameters are assumed to be linear, i.e., in terms of addition and multiplication by scalars. This is why we are restricted to *linear* optical systems. The qualifiers *quasi-monochromatic* and *plane* will be omitted from now on under the assumption that we are in fact dealing with this type of electromagnetic wave.

4.1 Propagation of light through anisotropic media

Changing the polarization state of light means modifying the coherency matrix elements, which in turn require that different components of the electric field vector are acted on differently by the medium. If E_x and E_y suffer the same alteration, a scaling of $\mathbf{C}$ is effected, so that the polarization state is unchanged. As a matter of fact, we have seen in the previous chapter how both the linear analyzer and the linear retarder act differently on given components of $\boldsymbol{E}$. The wave equation (2.1), however, predicts no different behavior for the orthogonal components. This apparent contradiction is resolved when we take into account that Eq. (2.1) is obtained for isotropic media, which are unable to change the polarization state of any light beam. Every pair of orthogonal components of the electric field propagates with the same velocity (as derived from the same dielectric permittivity) regardless of the direction. We are thus driven to consider anisotropic media. Optical anisotropy may be inherent to the medium because of its crystalline or molecular structure or may be induced by an external mechanism. Examples of the former type can be found in most laboratory devices like those we shall be discussing in Section 4.6. As far as this text is concerned, the paradigmatic case of the second type of anisotropic system is a stellar (solar) atmosphere permeated by a magnetic field (see Chapter 8).

If the medium presents dielectric permittivity anisotropies then the relationship between the electric displacement, $\boldsymbol{D}$, and the electric field, $\boldsymbol{E}$, is no longer a scaling relationship ($\boldsymbol{D} = \varepsilon \boldsymbol{E}$) and the wave equation does not hold as such. Now Maxwell's equations imply that $\boldsymbol{E}$ is not perpendicular to the wavefront normal, $\hat{\boldsymbol{s}}$, while $\boldsymbol{D}$ still is. It can be seen that $\boldsymbol{D}$ is proportional to $\boldsymbol{E}_{\perp}$, the vector component of $\boldsymbol{E}$ perpendicular to $\hat{\boldsymbol{s}}$ in the plane of $\boldsymbol{E}$ and $\hat{\boldsymbol{s}}$:

$$\boldsymbol{D} = \varepsilon \left[\boldsymbol{E} - (\boldsymbol{E} \cdot \hat{\boldsymbol{s}})\hat{\boldsymbol{s}} \right]. \tag{4.1}$$

Quite remarkably, the dielectric permittivity of Eq. (4.1) must depend on the wavefront direction. $\boldsymbol{D}$ and $\boldsymbol{E}$ then form an angle ϕ that is non-zero in general. Only when light propagates in certain directions within the medium does the scalar product of $\boldsymbol{E}$ and $\hat{\boldsymbol{s}}$ cancel out, and then $\boldsymbol{D}$ can be parallel to $\boldsymbol{E}$, as we shall see presently.

The magnetic field vector, $\boldsymbol{H}$, associated with the light perturbation remains perpendicular to both $\boldsymbol{D}$ and $\boldsymbol{E}$ and to the wave normal, $\hat{\boldsymbol{s}}$. The situation is illustrated in Fig. 4.1. If $\boldsymbol{D}$ and $\boldsymbol{E}$ are in the plane represented by the page, then $\boldsymbol{H}$ points out of the paper.

If we keep the Poynting vector, $\boldsymbol{S} = (c/8\pi)\langle \boldsymbol{E} \wedge \boldsymbol{H}^* \rangle$, as the vector for the energy flow then we are forced to conclude that *energy does not propagate along the wavefront normal* $\hat{\boldsymbol{s}}$ as in the case of isotropic media but along the *ray* direction $\hat{\boldsymbol{t}} = \boldsymbol{S}/S$ forming with $\hat{\boldsymbol{s}}$ the same angle ϕ as that between $\boldsymbol{D}$ and $\boldsymbol{E}$.

We must then distinguish between the phase velocity (or wavefront velocity), $v_{\mathrm{p}} = c/n$, and the ray velocity (or energy propagation velocity), $v_{\mathrm{r}} = v_{\mathrm{p}}/\cos\phi$. Note that both velocities vary for different directions $\hat{\boldsymbol{s}}$ or $\hat{\boldsymbol{t}}$. This is so because we can no longer talk of a single index of refraction for the medium. If the dielectric permittivity depends on direction, then so does the refractive index ($n = \sqrt{\varepsilon}$; we shall assume from now on that the magnetic permeability, μ, of the medium is unity, and that it is isotropic).

The simplest way of accounting for anisotropies in ε is to assume that the relation between $\boldsymbol{D}$ and $\boldsymbol{E}$ is linear;

$$\boldsymbol{D} = \boldsymbol{\varepsilon} \boldsymbol{E}, \tag{4.2}$$

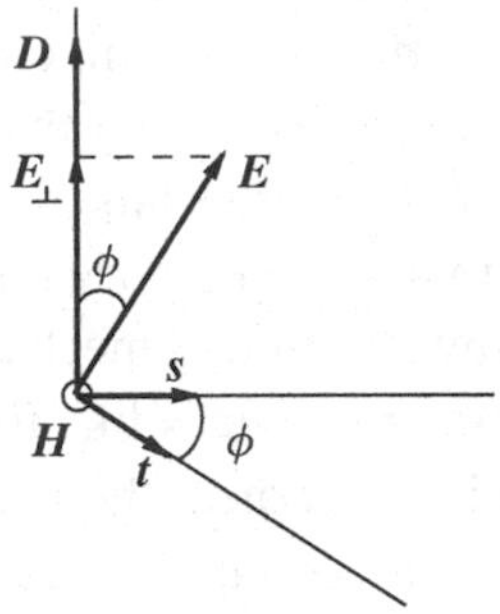

Fig. 4.1. Electric field vector, $\boldsymbol{E}$, electric displacement vector, $\boldsymbol{D}$, and magnetic field vector, $\boldsymbol{H}$ (pointing out of the paper), associated with the luminous disturbance in an anisotropic medium. $\boldsymbol{D}$ propagates in the direction of the wavefront normal, $\hat{\boldsymbol{s}}$, but energy propagates along the Poynting vector direction, $\hat{\boldsymbol{t}}$, the ray direction. Vectors $\hat{\boldsymbol{s}}$ and $\hat{\boldsymbol{t}}$ form the same angle as vectors $\boldsymbol{D}$ and $\boldsymbol{E}$.

where the dielectric permittivity is now a tensor (a 3×3 matrix). It can be shown, however, that relationship (4.2) simplifies in a given reference frame. The validity of the Poynting vector as the energy flow vector is a necessary and sufficient condition (e.g., Born and Wolf, 1993) for the dielectric tensor to be symmetric, which is of great help, for there always exists an orthogonal transformation ($\mathbf{T}\mathbf{T}^{\mathrm{T}} = \mathbf{T}^{\mathrm{T}}\mathbf{T} = \mathbb{1}$) that diagonalizes $\boldsymbol{\varepsilon}$:

$$\mathbf{T}\boldsymbol{\varepsilon}\mathbf{T}^{\mathrm{T}} = \boldsymbol{\varepsilon}' = \mathrm{diag}(\varepsilon_1, \varepsilon_2, \varepsilon_3), \tag{4.3}$$

where ε_1, ε_2, and ε_3 are the eigenvalues of the dielectric tensor.

If we change the $\{\hat{\boldsymbol{x}}, \hat{\boldsymbol{y}}, \hat{\boldsymbol{z}}\}$ reference frame by this transformation, we obtain the new frame $\{\hat{\boldsymbol{e}}_1, \hat{\boldsymbol{e}}_2, \hat{\boldsymbol{e}}_3 \mid \hat{\boldsymbol{e}}_i \cdot \hat{\boldsymbol{e}}_j = \delta_{i,j},\ i, j = 1, 2, 3\}$. This frame is called the *principal* reference frame and the unit vectors are oriented in the so-called *principal directions* of the medium. In such a reference frame, the relationship (4.2) becomes

$$\begin{pmatrix} D_1 \\ D_2 \\ D_3 \end{pmatrix} = \begin{pmatrix} \varepsilon_1 & 0 & 0 \\ 0 & \varepsilon_2 & 0 \\ 0 & 0 & \varepsilon_3 \end{pmatrix} \begin{pmatrix} E_1 \\ E_2 \\ E_3 \end{pmatrix}. \tag{4.4}$$

Therefore, in the frame of principal directions of the medium, the corresponding components of the electric displacement and of the electric field are proportional. If $\boldsymbol{E}$ oscillates in any of the principal directions, the angle ϕ between $\boldsymbol{D}$ and $\boldsymbol{E}$ is zero.

It can then be shown that the component equations of (2.1) keep the previous shape in the principal reference frame. The only difference is that each component now has its own dielectric permittivity:

$$\nabla^2 E_i - \frac{\varepsilon_i}{c^2}\ddot{E}_i = 0. \tag{4.5}$$

We thus have three main refractive indices, n_1, n_2, and n_3 ($n_i = \sqrt{\varepsilon_i}$), three principal phase velocities, $v_{\mathrm{p},1}$, $v_{\mathrm{p},2}$, and $v_{\mathrm{p},3}$ (c/n_i), and three principal ray velocities, $v_{\mathrm{r},1}$, $v_{\mathrm{r},2}$, and $v_{\mathrm{r},3}$. Any two orthogonal components of $\boldsymbol{E}$ (and of $\boldsymbol{D}$) propagate with different ray (and phase) velocities. This statement can be reformulated by saying that every direction, $\hat{\boldsymbol{s}}$, of the anisotropic medium admits the propagation of two waves linearly polarized in orthogonal directions. We now study both of them.

Depending on the values of the three principal permittivities (or refractive indices), two categories can be distinguished for classifying anisotropic media. Either two permittivities are equal but different to the third, or all three are different. Of course, if all the three are equal, we recover the isotropic medium case. The first class, for which, say, $\varepsilon_1 = \varepsilon_2 \neq \varepsilon_3$ and $v_1 = v_2 \neq v_3$ (both for the ray and phase velocities), contains the so-called *uniaxial* media. The second class contains media for which $\varepsilon_1 \neq \varepsilon_2 \neq \varepsilon_3$, so that $v_1 \neq v_2 \neq v_3$. Such media are called *biaxial*.

It is easily understood that uniaxial media present a given direction called the *optical axis* along which the two orthogonal components of the electric field stream with the same velocity. The optical axis is obviously $\hat{\boldsymbol{e}}_3$ in terms of the above convention. If $\hat{\boldsymbol{s}} = \hat{\boldsymbol{e}}_3$, then $\boldsymbol{E}$ has only components in the plane of $\hat{\boldsymbol{e}}_1$ and $\hat{\boldsymbol{e}}_2$. Since $\varepsilon_1 = \varepsilon_2$, both components travel with the same velocity, $v_1 = v_2$, and are parallel to the corresponding components of $\boldsymbol{D}$. In any other direction, the orthogonal components of $\boldsymbol{E}$ have different non-zero projections on $\hat{\boldsymbol{e}}_3$. Their propagation, then, should be studied separately.

Although it is not so easy to visualize, biaxial media present two optical axes, i.e., two directions for which the two orthogonal components of $\boldsymbol{E}$ stream with the same velocity. Since most laboratory devices have uniaxial properties, we shall restrict our analysis here to uniaxial media.† A magnetic stellar atmosphere indeed has three different refractive indices, so that it is optically biaxial (see Section 8.4).

For any given wavefront direction, $\hat{\boldsymbol{s}}$, the plane formed between the optical axis $\hat{\boldsymbol{e}}_3$ and $\hat{\boldsymbol{s}}$ is called the *principal plane* of the medium. The electric field vector and the electric displacement vector can both be considered as the sum of two vector components:

$$\begin{aligned} \boldsymbol{E} &= \boldsymbol{E}_{\mathrm{o}} + \boldsymbol{E}_{\mathrm{e}}, \\ \boldsymbol{D} &= \boldsymbol{D}_{\mathrm{o}} + \boldsymbol{D}_{\mathrm{e}}, \end{aligned} \tag{4.6}$$

where $\boldsymbol{E}_{\mathrm{e}}$ and $\boldsymbol{D}_{\mathrm{e}}$ are in the principal plane and $\boldsymbol{E}_{\mathrm{o}}$ and $\boldsymbol{D}_{\mathrm{o}}$ are in a plane perpendicular to this (see Fig. 4.2). This decomposition is convenient because, besides being colinear, $\boldsymbol{E}_{\mathrm{o}}$ and $\boldsymbol{D}_{\mathrm{o}}$ only have components in the plane of $\hat{\boldsymbol{e}}_1$ and $\hat{\boldsymbol{e}}_2$, regardless of $\hat{\boldsymbol{s}}$. It is readily seen that both propagate with a velocity $v_{\mathrm{o}} = v_1 = v_2$

† Although we have so far been speaking of the fast and the slow axes of a linear retarder, this does not mean that the device is biaxial. In fact, either of the axes can be the optical axis; the other simply marks the orthogonal direction of polarization.

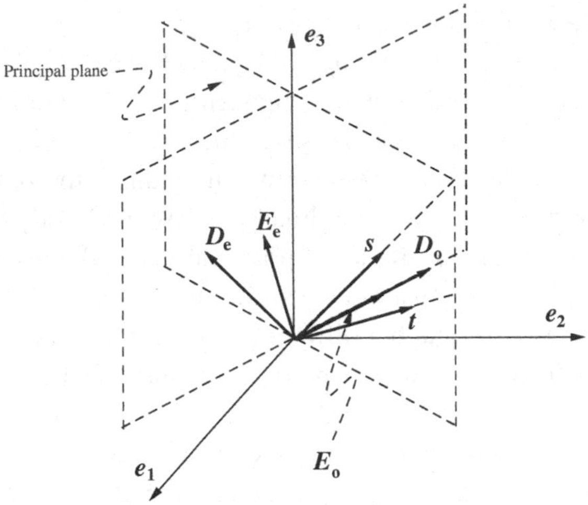

Fig. 4.2. The electric and displacement fields can always be decomposed into orthogonal components in the principal plane (the extraordinary components) and in a plane perpendicular to this (the ordinary components). Note that, necessarily, $D_\mathrm{o} \parallel E_\mathrm{o}$. Therefore, no distinction can be made between the wavefront direction and the ray direction for the ordinary component. In general, however, $D_\mathrm{e} \nparallel E_\mathrm{e}$.

independently of $\hat{\boldsymbol{s}}$. The velocity, v_e, of $\boldsymbol{E}_\mathrm{e}$, however, depends on the propagation direction since it has components in all three principal directions. The propagation of a light beam through an anisotropic medium can then be thought of as that of two linearly polarized beams, one having a velocity independent of direction, the *ordinary* beam, and the other with a velocity depending on direction, the *extraordinary* beam. The ordinary beam propagates "ordinarily", that is, like any beam through an isotropic medium: $\boldsymbol{E}_\mathrm{o} \parallel \boldsymbol{D}_\mathrm{o} \Longrightarrow \hat{\boldsymbol{t}} \parallel \hat{\boldsymbol{s}}$ and $v_{\mathrm{r,o}} = v_{\mathrm{p,o}}$. On the other hand, for the extraordinary beam $\hat{\boldsymbol{t}} \nparallel \hat{\boldsymbol{s}}$ and $v_{\mathrm{r,e}} \neq v_{\mathrm{p,e}}$, unless the wavefront direction is perpendicular to the optical axis. If $\hat{\boldsymbol{s}} \perp \hat{\boldsymbol{e}}_3$, then $\boldsymbol{E}_\mathrm{e} \parallel \boldsymbol{D}_\mathrm{e}$ also, because both oscillate along $\hat{\boldsymbol{e}}_3$. Note that a given uniaxial medium has as many extraordinary refractive indices as the possible directions for light propagation throughout the medium. By convention, manufacturers of polarization devices usually give just one ordinary, n_o, and one extraordinary, n_e, index. The latter is assumed to correspond to propagation perpendicular to $\hat{\boldsymbol{e}}_3$; that is, $n_\mathrm{e} = n_3$. In the following subsection we discuss the main observable consequences of the optically anisotropic structure of the medium.

4.1.1 Measurable effects of anisotropy

The first measurable effect of anisotropy is a phase difference between the ordinary and the extraordinary beams. Consider a light beam impinging on a uniaxial medium perpendicularly to the optical axis. In such a case, both the ordinary and extraordinary rays propagate in the direction of the wavefront normal: their electric fields and electric displacements are parallel. If n_o and n_e $(= n_3)$ are the ordinary and extraordinary refractive indices, one of the beams is retarded relative to the other with a linear dependence on the value of $\beta \equiv n_e - n_o$. If a slab of thickness t is traversed, then the total phase difference, called the *retardance*, is

$$\delta \equiv \frac{\beta t}{\lambda_0} = \frac{2\pi\beta t}{\lambda_0} \text{ rad,} \tag{4.7}$$

where λ_0 is the wavelength of the wave in vacuum. If $\beta > 0$, then the extraordinary beam is retarded by δ. The fast and slow axes we were referring to regarding retarders in Section 3.5 are evidently the optical axis and another direction perpendicular to this. Depending on the positiveness or negativeness of β, one or the other will coincide with the optical axis. Note that retardance depends on wavelength. Moreover, since n_e depends on direction, any other beam reaching the anisotropic medium in a direction other than that of a normal to $\hat{e}_3$ experiences a retardance between the components different from that given by Eq. (4.7). But this phenomenon is in fact connected with the second measurable effect.

The second effect is called *birefringence* or *double refraction*.† Snell's law of refraction applied to the transition from vacuum to the anisotropic material gives

$$\begin{aligned} \frac{\sin\alpha_i}{\sin\alpha_o} &= n_o; \\ \frac{\sin\alpha_i}{\sin\alpha_e} &= n_e(\hat{s}), \end{aligned} \tag{4.8}$$

so that the ordinary beam behaves "ordinarily" while the calculation of α_e is more complicated, since the extraordinary refractive index depends on direction (see the next section). In any case, the two beams necessarily split and behave independently within the medium. Note that double refraction is always accompanied by phase differences. If now t_o and t_e stand for the *geometric* paths of the ordinary and extraordinary rays, respectively, the retardance is given by

$$\delta = 2\pi \frac{n_e t_e - n_o t_o}{\lambda_0}. \tag{4.9}$$

A third effect called *dichroism* may appear in certain materials. It is just a difference between the absorption properties for orthogonal polarization states. As a consequence of dichroism, the ordinary beam may become more attenuated than the extraordinary one or vice versa.

† The parameter β is also called *the* birefringence of the medium, but for the effect under discussion there is no possible confusion.

A fourth effect may appear in some instances. Since $n_o \neq n_e$, total reflection can take place just for one of the beams at given interfaces between media. Recall that total reflection occurs at the boundary between an optically denser medium of refractive index n_a and an optically less dense medium of index n_b so that $n_b < n_a$. If α_a and α_b are the angles of the incident and refracted waves, Snell's law gives

$$\frac{\sin\alpha_a}{\sin\alpha_b} = \frac{n_b}{n_a} < 1. \tag{4.10}$$

Equation (4.10) provides no real solution for α_b when the incident angle is such that $\sin\alpha_a \geq n_b/n_a$. Hence, no refracted wave appears and the incident beam is totally reflected.

4.2 The extraordinary index of refraction and the energy propagation direction

As pointed out in the previous section, the extraordinary index of refraction depends on direction so that ray tracing in anisotropic media becomes somewhat elaborate.

Consider both $\boldsymbol{D}_e$ and $\boldsymbol{E}_e$ as the sum of two orthogonal components in the principal plane; namely,

$$\begin{aligned} \boldsymbol{D}_e &= \boldsymbol{D}_{e,3} + \boldsymbol{D}_{e,12}, \\ \boldsymbol{E}_e &= \boldsymbol{E}_{e,3} + \boldsymbol{E}_{e,12}, \end{aligned} \tag{4.11}$$

where $\boldsymbol{D}_{e,3}$ and $\boldsymbol{E}_{e,3}$ are in the $\hat{\boldsymbol{e}}_3$ direction, and $\boldsymbol{D}_{e,12}$ and $\boldsymbol{E}_{e,12}$ are in the plane of $\hat{\boldsymbol{e}}_1$ and $\hat{\boldsymbol{e}}_2$ (i.e., $\boldsymbol{D}_{e,12} = \boldsymbol{D}_{e,1} + \boldsymbol{D}_{e,2}$; $\boldsymbol{E}_{e,12} = \boldsymbol{E}_{e,1} + \boldsymbol{E}_{e,2}$).

Equation (4.4) implies that

$$\boldsymbol{D}_{e,3} = \varepsilon_3 \boldsymbol{E}_{e,3}, \tag{4.12}$$

and that

$$\boldsymbol{D}_{e,12} = \varepsilon_o \boldsymbol{E}_{e,12}, \tag{4.13}$$

where, obviously, $\varepsilon_o = \varepsilon_1 = \varepsilon_2$.

Now $\varepsilon_e(\hat{\boldsymbol{s}})$ represents the proportionality constant between $\boldsymbol{D}_e$ and $\boldsymbol{E}_{e,\perp}$, the component of $\boldsymbol{E}_e$ perpendicular to $\hat{\boldsymbol{s}}$ [Eq. (4.1)]:

$$\boldsymbol{D}_e = \varepsilon_e(\hat{\boldsymbol{s}}) \left[\boldsymbol{E}_e - (\boldsymbol{E}_e \cdot \hat{\boldsymbol{s}})\hat{\boldsymbol{s}}\right]. \tag{4.14}$$

With a little algebra, using Eqs (4.12) and (4.13) and projecting $\boldsymbol{D}_e$ from Eq. (4.14), the vector components of the electric displacement turn out to be

$$\boldsymbol{D}_{e,3} = -\frac{\boldsymbol{E}_e \cdot \hat{\boldsymbol{s}}}{[(1/\varepsilon_e) - (1/\varepsilon_3)]} s_3 \tag{4.15}$$

and

$$\boldsymbol{D}_{\mathrm{e},12} = -\frac{\boldsymbol{E}_{\mathrm{e}} \cdot \hat{\boldsymbol{s}}}{[(1/\varepsilon_{\mathrm{e}}) - (1/\varepsilon_{\mathrm{o}})]} \boldsymbol{s}_{12}, \tag{4.16}$$

where $\boldsymbol{s}_3$ and $\boldsymbol{s}_{12}$ are the vector components of $\hat{\boldsymbol{s}}$ along $\hat{\boldsymbol{e}}_3$ and in the plane of $\hat{\boldsymbol{e}}_1$ and $\hat{\boldsymbol{e}}_2$ $(\boldsymbol{s}_1 + \boldsymbol{s}_2)$.

Since $\boldsymbol{D}_{\mathrm{e}} \perp \hat{\boldsymbol{s}}$, Eqs (4.11), (4.15), and (4.16) give

$$\frac{s_3^2}{[(1/\varepsilon_{\mathrm{e}}) - (1/\varepsilon_3)]} + \frac{s_{12}^2}{[(1/\varepsilon_{\mathrm{e}}) - (1/\varepsilon_{\mathrm{o}})]} = 0. \tag{4.17}$$

Let γ be the angle between $\hat{\boldsymbol{s}}$ and $\hat{\boldsymbol{e}}_3$. Equation (4.17) can then be recast in the form

$$\frac{1}{\varepsilon_{\mathrm{e}}(\hat{\boldsymbol{s}})} = \frac{1}{\varepsilon_{\mathrm{o}}} \cos^2 \gamma + \frac{1}{\varepsilon_3} \sin^2 \gamma, \tag{4.18}$$

and, remembering the definition of refractive index [Eq. (2.2)],

$$\frac{1}{n_{\mathrm{e}}^2(\gamma)} = \frac{1}{n_{\mathrm{o}}^2} \cos^2 \gamma + \frac{1}{n_3^2} \sin^2 \gamma, \tag{4.19}$$

where we have assumed unit magnetic permeability for the medium, which is the case for most materials of interest to us. Then, $n_{\mathrm{e}}(\gamma)$ can be interpreted as *the* refractive index for the extraordinary beam, which, it should be noted, depends only on the angle between the wavefront normal and the optical axis. In the two limiting cases of $\gamma = 0$ and $\gamma = \pi/2$ the refractive index for the extraordinary beam behaves as expected: $n_{\mathrm{e}}(0) = n_{\mathrm{o}}$ and $n_{\mathrm{e}}(\pi/2) = n_3$.

Once we know $n_{\mathrm{e}}(\gamma)$, Snell's law [Eq. (4.8)] provides the direction of the refracted wave normal. Let us now calculate the ray direction, $\hat{\boldsymbol{t}}$, or, in other words, the angle θ between $\hat{\boldsymbol{t}}$ and $\hat{\boldsymbol{e}}_3$. (Recall that $\hat{\boldsymbol{t}}$ is also in the principal plane.) Since $D_{\mathrm{e},3} = D_{\mathrm{e}} \sin \gamma$ and $D_{\mathrm{e},12} = D_{\mathrm{e}} \cos \gamma$, Eqs (4.12) and (4.13) imply that

$$E_{\mathrm{e},3} = \frac{1}{\varepsilon_3} D_{\mathrm{e}} \sin \gamma, \tag{4.20}$$

and that

$$E_{\mathrm{e},12} = \frac{1}{\varepsilon_{\mathrm{o}}} D_{\mathrm{e}} \cos \gamma. \tag{4.21}$$

On the other hand we have $E_{\mathrm{e},3} = E_{\mathrm{e}} \sin \theta$ and $E_{\mathrm{e},12} = E_{\mathrm{e}} \cos \theta$. Hence, we obtain

$$\sin \theta = \frac{1}{\varepsilon_3} \frac{D_{\mathrm{e}}}{E_{\mathrm{e}}} \sin \gamma \tag{4.22}$$

and

$$\cos \theta = \frac{1}{\varepsilon_{\mathrm{o}}} \frac{D_{\mathrm{e}}}{E_{\mathrm{e}}} \cos \gamma, \tag{4.23}$$

from which we finally derive:

$$\tan\theta = \frac{n_o^2}{n_3^2}\tan\gamma. \tag{4.24}$$

Therefore, if $n_o < n_3$ (positive birefringence) then $\theta < \gamma$ ($\theta = \gamma - \phi$). If the birefringence of the medium is negative, then $\theta > \gamma$ ($\theta = \gamma + \phi$).

4.3 Some notational conventions

Before entering into details, it is convenient to clearly establish a few notational conventions that are useful in practice and customary in polarization optics. These conventions, however, are not widely used in the astrophysical literature. By default, vectors are assumed to be column vectors, hence the above notation $\boldsymbol{I} = (I, Q, U, V)^{\mathrm{T}}$ for the Stokes vector. Four-vectors may be denoted by a scalar followed by a three-dimensional vector. With such a convention, the Stokes vector can be written as

$$\boldsymbol{I} = (I, Q, U, V)^{\mathrm{T}} = I(1, \boldsymbol{p}^{\mathrm{T}})^{\mathrm{T}}. \tag{4.25}$$

Formally, 4 × 4 matrices may be written as 2 × 2 matrices in which the first element is a scalar, the second element of the first row is a (transposed) three-vector, the first element of the second row is a three-vector, and the second element of the second row is a 3 × 3 matrix. Therefore, a general 4 × 4 matrix $\mathbf{A}$ may be written as

$$\mathbf{A} = \begin{pmatrix} a & \boldsymbol{b}^{\mathrm{T}} \\ \boldsymbol{c} & \mathbf{D} \end{pmatrix} \equiv \begin{pmatrix} a_{00} & a_{01} & a_{02} & a_{03} \\ a_{10} & a_{11} & a_{12} & a_{13} \\ a_{20} & a_{21} & a_{22} & a_{23} \\ a_{30} & a_{31} & a_{32} & a_{33} \end{pmatrix}, \tag{4.26}$$

whence

$$\begin{aligned} a &= a_{00}, \\ \boldsymbol{b} &= (a_{01}, a_{02}, a_{03})^{\mathrm{T}}, \\ \boldsymbol{c} &= (a_{10}, a_{20}, a_{30})^{\mathrm{T}}, \\ \mathbf{D} &= \begin{pmatrix} a_{11} & a_{12} & a_{13} \\ a_{21} & a_{22} & a_{23} \\ a_{31} & a_{32} & a_{33} \end{pmatrix}. \end{aligned} \tag{4.27}$$

Finally, products are understood as products between matrices. Symbols are omitted and $\boldsymbol{a}^{\mathrm{T}}\boldsymbol{b}$ is understood as the scalar product and $\boldsymbol{a}\boldsymbol{b}^{\mathrm{T}}$ is understood as the tensor product (i.e., a matrix) between vectors $\boldsymbol{a}$ and $\boldsymbol{b}$.

The reasons for such a notation are at least four-fold: (1) the total intensity of a light beam is considered as a scaling factor; (2) the polarization vector is explicitly used, hence stressing that it belongs to the Poincaré sphere; (3) since $\boldsymbol{p}$ must belong to the Poincaré sphere, transformations of the polarization state of light are seen as transformations of three-vectors, i.e., geometrical rotations, contractions, or dilatations; (4) the algebraic manipulations of $\mathbb{R}^4$ elements and transformations are simplified to those of $\mathbb{R}^3$,which are easier to handle.

4.4 Transforming the polarization state of light

Since every linear transformation in $\mathbb{R}^4$ can be described by a 4×4 matrix, any linear change in the polarization state of light is represented by a given 4×4 matrix, the so-called Mueller matrix,

$$\boldsymbol{I}' = \mathbf{M}\boldsymbol{I}, \tag{4.28}$$

or, more specifically,

$$\begin{pmatrix} I' \\ Q' \\ U' \\ V' \end{pmatrix} = \begin{pmatrix} M_{00} & M_{01} & M_{02} & M_{03} \\ M_{10} & M_{11} & M_{12} & M_{13} \\ M_{20} & M_{21} & M_{22} & M_{23} \\ M_{30} & M_{31} & M_{32} & M_{33} \end{pmatrix} \begin{pmatrix} I \\ Q \\ U \\ V \end{pmatrix}, \tag{4.29}$$

and, in block component notation:

$$g \begin{pmatrix} 1 \\ \boldsymbol{p}' \end{pmatrix} = M_{00} \begin{pmatrix} 1 & \boldsymbol{h}^{\mathrm{T}} \\ \boldsymbol{v} & \mathbf{N} \end{pmatrix} \begin{pmatrix} 1 \\ \boldsymbol{p} \end{pmatrix}, \tag{4.30}$$

where g is the *gain* or *transmittance* of the system, that is, the ratio between output and input intensities,

$$g \equiv \frac{I'}{I} \geq 0, \tag{4.31}$$

and the identification of $\boldsymbol{h}$, $\boldsymbol{v}$, and $\mathbf{N}$ with the matrix elements of Eq. (4.29) is straightforward after Eq. (4.27). Note that in fact $g = g(\boldsymbol{p})$; i.e., the transmittance depends on the input polarization state.

4.5 The Mueller matrix and some of its properties

Not every four-vector is a physically meaningful Stokes vector. Conditions (3.19) and (3.20) must be fulfilled. Consequently, not every 4×4 matrix can be a physically meaningful Mueller matrix. To meet this definition a 4×4 must transform physical Stokes vectors onto physical Stokes vectors or, equivalently, Poincaré vectors onto Poincaré vectors, as illustrated in Fig. 4.3. This physical condition

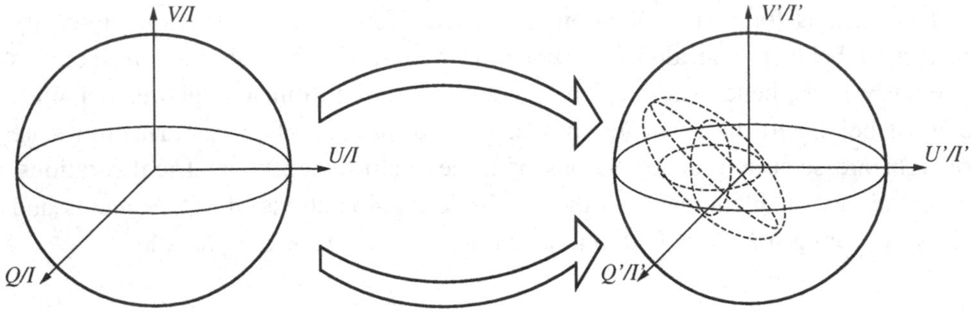

Fig. 4.3. A linear optical system maps the Poincaré sphere onto a subset (an ellipsoid) of the Poincaré sphere.

requires the Mueller matrix elements to have several properties, some of which deserve explicit mention:

- The first element must be non-negative:

$$M_{00} \geq 0. \tag{4.32}$$

  This property holds because M_{00} turns out to be the output intensity corresponding to a natural input. This is easy to see from Eq. (4.29) once it is recalled that natural light has $Q = U = V = 0$ (Section 3.4).
- The first row of a physically meaningful Mueller matrix is a physically meaningful Stokes vector:

$$M_{00} \geq \sqrt{M_{01}^2 + M_{02}^2 + M_{03}^2}. \tag{4.33}$$

  In other words, the vector $\boldsymbol{h}$ appearing in Eq. (4.30) belongs to the Poincaré sphere and is called the *diattenuation vector*. Condition (4.33) is equivalent (necessary and sufficient) to the ratio of the output to input intensities being positive (i.e., $I'/I \geq 0$). Let us first demonstrate the necessary part of the statement.

  Assume that inequality (4.33) is true. Since, according to Eq. (3.22),

$$\sum_{i=1}^{3} p_i^2 \leq 1,$$

  then

$$M_{00}^2 \geq \sum_{i=1}^{3} M_{0i}^2 \sum_{i=1}^{3} p_i^2,$$

which, according to the Schwarz's inequality, gives

$$M_{00}^2 \geq \left(\sum_{i=1}^{3} M_{0i} p_i \right)^2 ;$$

whence

$$M_{00} \geq \left| \sum_{i=1}^{3} M_{0i} p_i \right| \geq - \sum_{i=1}^{3} M_{0i} p_i .$$

Therefore, the first of Eqs (4.29) yields

$$\frac{I'}{I} = M_{00} + \sum_{i=1}^{3} M_{0i} p_i \geq 0.$$

For the sufficient part we proceed as follows. We look for a minimum of

$$f(p_i) \equiv \sum_{i=1}^{3} M_{0i} p_i .$$

Since there is no absolute minimum of f inside (or on the surface of) the Poincaré sphere, we look for a minimum conditioned to be at the surface of $\mathbb{P}$. Let l be defined as

$$l(p_i) \equiv f(p_i) - k \left(\sum_{i=1}^{3} p_i^2 - 1 \right),$$

where k is a Lagrange multiplier. The minimum of l is located at

$$p_{i,\min} = \frac{M_{0i}}{2k},$$

provided that the second derivative is positive; that is, provided that $k < 0$. Since the minimum is on $\mathbb{P}$, $\sum_{i=1}^{3} p_{i,\min}^2 = 1$, whence

$$2k = - \sqrt{\sum_{i=1}^{3} M_{0i}^2} .$$

Therefore, the minimum of f we were looking for is

$$f_{\min} \equiv f(p_{i,\min}) = - \sqrt{\sum_{i=1}^{3} M_{0i}^2} .$$

Taking into account now that $I'/I = M_{00}+f(p_i) \geq 0$ for every p_i, we obviously have:

$$M_{00} \geq \sqrt{\sum_{i=1}^{3} M_{0i}^2}$$

as we sought to demonstrate.

- The first column of a physically meaningful Mueller matrix is a physically meaningful Stokes vector:

$$M_{00} \geq \sqrt{M_{10}^2 + M_{20}^2 + M_{30}^2}. \tag{4.34}$$

  This inequality is obtained after bearing in mind that the first column is the output corresponding to a natural input. Vector $\boldsymbol{v}$ in Eq. (4.30) also belongs to the Poincaré sphere and is called the *polarizance vector*.

The diattenuation vector combines input polarization into output intensity and is seen to govern the gain of the system as a function of the input polarization: the first row of Eq. (4.30) implies that

$$g(\boldsymbol{p}) = M_{00}(1 + \boldsymbol{h}^{\mathrm{T}}\boldsymbol{p}). \tag{4.35}$$

The transmittance, then, is not independent of the input unless $\boldsymbol{h} = \boldsymbol{0}$. Equations (4.35) and (4.31) readily explain why M_{00} must be non-negative [Eq. (4.32)]: if the input is natural then $\boldsymbol{p} = \boldsymbol{0}$ and M_{00} equals the (non-negative) gain. One can also obtain maximum and minimum transmittance from Eq. (4.35):

$$1 - h \leq \frac{g}{M_{00}} \leq 1 + h; \tag{4.36}$$

in other words, g can be neither smaller than 0 nor greater than $2M_{00}$ for any linear optical system. Consequently, passive optical systems (those that do not have emission properties) must not have M_{00} greater than 0.5 unless they have a zero diattenuation vector, since the gain cannot be greater than unity.

The polarizance vector contributes the input intensity to the output polarization. The second row of Eq. (4.30) gives

$$\frac{g}{M_{00}}\boldsymbol{p}' = \boldsymbol{v} + \mathbf{N}\boldsymbol{p}. \tag{4.37}$$

Hence, if $\boldsymbol{v} = \boldsymbol{0}$, natural light on input remains natural on output. In fact, it can be shown that a system whose polarizance vector is zero has no *empolarizing* capabilities, i.e., it cannot increase the degree of polarization of the incoming beam (see Lu and Chipman, 1998). This last property will be of importance when analyzing the elements of the propagation matrix of an anisotropic medium (Section 7.4).

Equation (4.37) also tells us that matrix **N** blends linear with linear and linear with circular polarizations and vice versa. It plays the role of a rotator and/or a dilator/contractor in the Poincaré sphere.

All the properties described so far concerning the Mueller matrix elements, and many others that have been omitted, are necessary conditions. Surprisingly,† a necessary and sufficient condition for a given 4×4 matrix to be a physically meaningful Mueller matrix was not obtained until 1993 by Givens and Kostinski. An alternative formulation of the same condition, suitable for application to experimental Mueller matrices, can be found in Landi Degl'Innocenti and del Toro Iniesta (1998).

4.6 Block components of (solar) polarimeters

As outlined in Section 3.5, measuring the polarization state of light relies upon transforming that state by means of specially suited devices, (compound) analyzers, also called *polarimeters*. The transformation is such that we measure intensities (hereafter also referred to as polarimetric signals) that are linear combinations of the Stokes parameters [e.g., Eq. (3.44)]. Once we know the basic mathematics of such a transformation, it is convenient to characterize those block components of the polarimeter by explicitly giving their Mueller matrices, which are obtained either by applying Eq. (4.29) to particular inputs for which the output is known or by direct derivation from the characteristic changes of the vector electric field after traversing the system.

4.6.1 Rotation of the reference frame

Since the reference frame is very relevant in describing the vector properties of the electric field and is in fact governed by the properties of the medium, let us first consider possible rotations in a plane perpendicular to the propagation direction. Every such rotation of angle θ measured counter-clockwise from the positive X axis is described by the matrix

$$\mathbf{G}(\theta) = \begin{pmatrix} 1 & 0 & 0 & 0 \\ 0 & c_2 & s_2 & 0 \\ 0 & -s_2 & c_2 & 0 \\ 0 & 0 & 0 & 1 \end{pmatrix} \tag{4.38}$$

such that

$$\mathbf{G}\mathbf{G}^{\mathrm{T}} = \mathbf{G}^{\mathrm{T}}\mathbf{G} = \mathbb{1}, \tag{4.39}$$

† The Mueller matrix formalism was proposed in the 1940s.

where $\mathbf{G}^{\mathrm{T}}(\theta) = \mathbf{G}(-\theta)$, $\mathbb{1}$ is the 4×4 identity matrix, and c_2 and s_2 are $\cos 2\theta$ and $\sin 2\theta$, respectively.

After a rotation of the reference frame, Stokes vectors and Mueller matrices transform according to the following formulae:

$$\begin{aligned} \boldsymbol{I}' &= \mathbf{G}^{\mathrm{T}}\boldsymbol{I}; \\ \mathbf{M}' &= \mathbf{G}^{\mathrm{T}}\mathbf{M}\mathbf{G}. \end{aligned} \tag{4.40}$$

Note that the rotation angle is doubled because of the quadratic nature of the Stokes parameters as functions of the Cartesian components of the electric field. Moreover, a rotation of angle θ of the reference frame is felt by the Stokes vector as a rotation of $-\theta$. The reason is simple: linearly polarized light at an angle α becomes linearly polarized at an angle $\alpha - \theta$ after rotation of the frame.

The shape of matrix $\mathbf{G}$ directly implies that both I and V are independent of the reference system, while Q and U are not. A rotation of the reference system changes the values of the linear polarization parameters but $Q^2 + U^2$ remains unaltered. In other words, the intrinsic parameters of the light beam are the intensity and the degrees of linear polarization, $\sqrt{Q^2 + U^2}/I$, and of circular polarization, V/I. Of course, the total degree of polarization, p, is also independent of the reference system.

4.6.2 The linear analyzer–polarizer

Analyzers are those polarization devices used for deriving the polarization state of light. Since we measure only intensities, the analyzers could be characterized by transmittance properties dependent on the input polarization state. In this broad sense, every linear system with a non-zero diattenuation vector could act as an analyzer in the sense that it always presents (different) maximum and minimum transmittance [Eq. (4.36)]. The maximum transmittance, $M_{00}(1 + h)$, is obtained for vectors $\boldsymbol{p}$ forming an angle $\arccos(1/p)$ with the diattenuation vector, and the minimum transmittance, $M_{00}(1 - h)$, is reached for $-\boldsymbol{p}$. That is, maximum transmittance is reached when $\boldsymbol{h}^{\mathrm{T}}\boldsymbol{p} = h$ and minimum transmittance results for a polarization vector such that $\boldsymbol{h}^{\mathrm{T}}\boldsymbol{p} = -h$. As illustrated in Fig. 4.4, all vectors lying within or on the surface of the upper cone of the figure form an angle $\arccos(1/p)$ with $\boldsymbol{h}$. Those lying within or on the surface of the lower cone form an angle $\arccos(-1/p)$ with $\boldsymbol{h}$. Note that all these vectors necessarily have magnitudes $p \geq 1$ since the base of the cone is tangential at just one point to $\mathbb{P}$. Vectors of magnitude $p > 1$ do not belong to $\mathbb{P}$. Thus, the only vector of the cone belonging to $\mathbb{P}$ is $\hat{\boldsymbol{h}}$. Depending on the input degree of polarization, the cone is broader or narrower, but qualitatively similar: all the vectors within it except $\hat{\boldsymbol{h}}$ have magnitudes

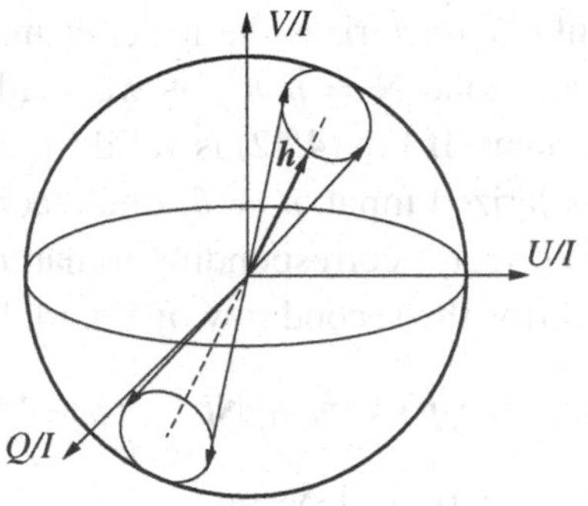

Fig. 4.4. Every Mueller matrix with diattenuation vector $\boldsymbol{h}$ has maximum transmittance for those vectors forming an angle $\arccos(1/p)$ (upper cone) and minimum transmittance for those forming an angle $\arccos(-1/p)$ (lower cone). Among these, the only physically meaningful polarization vectors are $\hat{\boldsymbol{h}}$ and $-\hat{\boldsymbol{h}}$ because the other have $p > 1$. The dashed line represents a diameter of the Poincaré sphere.

greater than unity. Therefore, the only physically meaningful polarization state for which transmittance is maximum turns out to be $\hat{\boldsymbol{h}}$. Analogously, transmittance is at a minimum for $-\hat{\boldsymbol{h}}$. If the diattenuation vector is already a unit vector, the maximum transmittance is $2M_{00}$ and the minimum transmittance is zero. These arguments lead to the definition of analyzers that follows.

A linear optical system whose transmittance is $2M_{00}$ for a given totally polarized state $\hat{\boldsymbol{p}}_0$ and completely blocks the orthogonal state $-\hat{\boldsymbol{p}}_0$ is called a *perfect analyzer*, or simply an analyzer for $\hat{\boldsymbol{p}}_0$. In mathematical terms, this definition leads to

$$\begin{aligned} g_{\max} &= g(\hat{\boldsymbol{p}}_0) = 2M_{00}; \\ g_{\min} &= g(-\hat{\boldsymbol{p}}_0) = 0. \end{aligned} \tag{4.41}$$

According to Eqs (4.35) and (4.36), conditions (4.41) are fulfilled if and only if $\boldsymbol{h} = \hat{\boldsymbol{p}}_0$. Since $h = p_0 = 1$, we can then say that a (physically meaningful) Mueller matrix describes a perfect analyzer if and only if it has a unit diattenuation vector $\hat{\boldsymbol{h}}$. It then describes an $\hat{\boldsymbol{h}}$ analyzer.

Rather than regarding the input polarization state as in the case of analyzers, one may be interested in characterizing some polarization systems by attending to the properties of the outcoming state of polarization. The definition of polarizers follows the latter option.

A linear optical system whose output is always a given totally polarized vector $\hat{\boldsymbol{p}}_0$ is called a *perfect polarizer*, or simply a $\hat{\boldsymbol{p}}_0$ polarizer. For such a system, Eq. (4.30) becomes

$$\frac{g(\boldsymbol{p})}{M_{00}} \begin{pmatrix} 1 \\ \hat{\boldsymbol{p}}_0 \end{pmatrix} = \begin{pmatrix} 1 & \boldsymbol{h}^{\mathrm{T}} \\ \boldsymbol{v} & \mathbf{N} \end{pmatrix} \begin{pmatrix} 1 \\ \boldsymbol{p} \end{pmatrix}, \quad \forall \boldsymbol{p} \in \mathbb{P}. \tag{4.42}$$

Equation (4.42) certainly characterizes the Mueller matrix of the system since it holds if and only if $\boldsymbol{v} = \hat{\boldsymbol{p}}_0$ and $\mathbf{N} = \hat{\boldsymbol{p}}_0 \boldsymbol{h}^{\mathrm{T}}$, as we shall see. First, consider the necessary part of the statement. If Eq. (4.42) is valid for all polarization states, for the particular case of unpolarized input, $\boldsymbol{p} = \boldsymbol{0}$, one readily obtains $\boldsymbol{v} = \hat{\boldsymbol{p}}_0$, since $(1, \boldsymbol{v}^{\mathrm{T}})^{\mathrm{T}}$ is the output Stokes vector corresponding to natural light for *every* optical system. Moreover, necessarily, the second row of Eq. (4.42) tells us that

$$\hat{\boldsymbol{p}}_0(1 + \boldsymbol{h}^{\mathrm{T}}\boldsymbol{p}) = \hat{\boldsymbol{p}}_0 + \mathbf{N}\boldsymbol{p}, \quad \forall \boldsymbol{p} \in \mathbb{P}, \tag{4.43}$$

since Eq. (4.35) holds for every optical system.

Solving Eq. (4.43) for $\hat{\boldsymbol{p}}_0$ we obtain

$$\hat{\boldsymbol{p}}_0 = \frac{\mathbf{N}\boldsymbol{p}}{\boldsymbol{h}^{\mathrm{T}}\boldsymbol{p}}, \quad \forall \boldsymbol{p} \in \mathbb{P}, \tag{4.44}$$

but this can happen if and only if

$$\mathbf{N} = \hat{\boldsymbol{p}}_0 \boldsymbol{h}^{\mathrm{T}}. \tag{4.45}$$

This last result (4.45) provides the way back to demonstrating the sufficient part of the above statement. Therefore, the general structure of a polarizer Mueller matrix is

$$\mathbf{M} = M_{00} \begin{pmatrix} 1 & \boldsymbol{h}^{\mathrm{T}} \\ \hat{\boldsymbol{v}} & \hat{\boldsymbol{v}}\boldsymbol{h}^{\mathrm{T}} \end{pmatrix}. \tag{4.46}$$

It is interesting to note that, if $\mathbf{M}$ is the Mueller matrix of a polarizer, then its transpose, $\mathbf{M}^{\mathrm{T}}$, is necessarily the Mueller matrix of an analyzer. Lu and Chipman (1996) say that the condition is also sufficient but give no convincing demonstration. Obviously, if $\boldsymbol{h}$ is also a unit vector in Eq. (4.46), the system is both a perfect polarizer and a perfect analyzer.

Particularizing to the case of our *linear* analyzer–polarizer (as defined in Section 3.5), it is easy to show that

$$\mathbf{L}(\theta) = \frac{1}{2} \begin{pmatrix} 1 & c_2 & s_2 & 0 \\ c_2 & c_2^2 & c_2 s_2 & 0 \\ s_2 & c_2 s_2 & s_2^2 & 0 \\ 0 & 0 & 0 & 0 \end{pmatrix}, \tag{4.47}$$

where $c_2 = \cos 2\theta$, $s_2 = \sin 2\theta$, and θ is the angle of the optical axis relative to the positive X axis.

A glance at Mueller matrix (4.47) readily indicates that the linear analyzer–polarizer always increases to unity the degree of polarization of partially polarized light because light is always linearly polarized on output at an angle θ: $\boldsymbol{I}' = 1/2(I + c_2 Q + s_2 U)(1, c_2, s_2, 0)^{\mathrm{T}}$. Moreover, rotating the device implies a modulation of the polarimetric signal according to the variation of the elements of its diattenuation vector.

4.6.3 The partial linear polarizer

The linear analyzer–polarizer described in the previous section is a particular case of the system we are dealing with now. Imagine that, instead of showing maximum transmittance and completely blocking two orthogonal components of the electric field vector, the amplitudes E_θ and $E_{\theta+90^\circ}$ were affected by two positive constant factors, k_1 and k_2, respectively. If we denote $\alpha \equiv k_1^2 + k_2^2$, $\beta \equiv (k_1^2 - k_2^2)/\alpha$, and $\gamma \equiv 2k_1k_2/\alpha$, the Mueller matrix of such a system turns out to be

$$\mathbf{M}(\theta) = \frac{\alpha}{2} \begin{pmatrix} 1 & \beta c_2 & \beta s_2 & 0 \\ \beta c_2 & c_2^2 + \gamma s_2^2 & (1-\gamma)c_2 s_2 & 0 \\ \beta s_2 & (1-\gamma)c_2 s_2 & s_2^2 + \gamma c_2^2 & 0 \\ 0 & 0 & 0 & \gamma \end{pmatrix}. \tag{4.48}$$

Note that matrix $\mathbf{M}(\theta)$ reduces to $\mathbf{L}(\theta)$ in Eq. (4.47) when $k_1 = 1$ and $k_2 = 0$. Equation (4.48) describes the Mueller matrix of a *partial linear polarizer*. The name comes from the fact that it outputs the partially polarized state, $\boldsymbol{p}' = 1/2(k_1^2 - k_2^2)\,(c_2, s_2, 0)^{\mathrm{T}}$, as a response to an unpolarized input. Nevertheless, the exiting polarization state depends on the input state. For instance, the system delivers elliptical (partially) polarized light for a circular (totally) polarized input.

4.6.4 The linear retarder

A linear retarder of retardance δ with its fast axis at an angle θ with the X axis has a Mueller matrix of the form

$$\mathbf{R}(\theta, \delta) = \begin{pmatrix} 1 & 0 & 0 & 0 \\ 0 & c_2^2 + s_2^2 \cos\delta & c_2 s_2(1-\cos\delta) & -s_2 \sin\delta \\ 0 & c_2 s_2(1-\cos\delta) & s_2^2 + c_2^2 \cos\delta & c_2 \sin\delta \\ 0 & s_2 \sin\delta & -c_2 \sin\delta & \cos\delta \end{pmatrix}. \tag{4.49}$$

Therefore, the linear retarder neither increases the degree of polarization nor modulates the polarimetric signal because of its zero polarizance and diattenuation vectors. It only rotates the input polarization vector in the Poincaré sphere since its 3×3 $\mathbf{N}$ matrix is a rotation matrix in $\mathbb{R}^3$. This property can easily be checked by verifying that $\mathbf{N}\mathbf{N}^{\mathrm{T}} = \mathbf{N}^{\mathrm{T}}\mathbf{N} = \mathbb{1}$ and $\det(\mathbf{N}) = 1$. Retarders are also called *wave plates*. When $\delta = \pi/2$ rad, i.e., $\beta t = \lambda_0/4$ for input light perpendicular to the optical axis [Eq. (4.7)], the retarder is said to be a *quarter-wave plate*; when $\delta = \pi$ rad, the retarder is called a *half-wave plate*.

4.6.5 The Mueller matrix of an optical train

In most instances, the polarimeter is made up of various components as for the case of the prototypical polarimeter we have been discussing. The total transformation is then described by the product of the individual Mueller matrices in the proper order, since matrix multiplication is non-commutative. If light encounters n block components, with $i = 1, 2, \ldots, n$ in this order, each one having a Mueller matrix $\mathbf{M}_i$, the Mueller matrix of the whole system is

$$\mathbf{M} = \mathbf{M}_n \mathbf{M}_{n-1} \ldots \mathbf{M}_2 \mathbf{M}_1. \tag{4.50}$$

It is worth noting that the Mueller matrix of two (or more) linear retarders is the Mueller matrix of an equivalent retarder, because the product of $\mathbb{R}^3$ rotation matrices is itself a rotation matrix:

$$\mathbf{R}_1 \mathbf{R}_2 = \begin{pmatrix} 1 & \boldsymbol{0}^{\mathrm{T}} \\ \boldsymbol{0} & \mathbf{N}_1 \end{pmatrix} \begin{pmatrix} 1 & \boldsymbol{0}^{\mathrm{T}} \\ \boldsymbol{0} & \mathbf{N}_2 \end{pmatrix} = \begin{pmatrix} 1 & \boldsymbol{0}^{\mathrm{T}} \\ \boldsymbol{0} & \mathbf{N}_1 \mathbf{N}_2 \end{pmatrix}. \tag{4.51}$$

This mathematical property very nicely fits the physical action of retarders on the electric field vector: two successive phase shifts, δ_1 and δ_2, must be equivalent to a single shift, $\delta_1 + \delta_2$.

The Mueller matrix of our example (see Fig. 4.5), a retarder whose fast axis is inclined at an angle θ plus a linear polarizer at 0°, is $\mathbf{M}(\theta, \delta) = \mathbf{L}(0)\mathbf{R}(\theta, \delta)$:

$$\mathbf{M}(\theta, \delta) = \frac{1}{2} \begin{pmatrix} 1 & c_2^2 + s_2^2 \cos\delta & c_2 s_2 (1 - \cos\delta) & -s_2 \sin\delta \\ 1 & c_2^2 + s_2^2 \cos\delta & c_2 s_2 (1 - \cos\delta) & -s_2 \sin\delta \\ 0 & 0 & 0 & 0 \\ 0 & 0 & 0 & 0 \end{pmatrix}. \tag{4.52}$$

Note that $\mathbf{M}$ is the Mueller matrix of an analyzer and of a polarizer ($h = 1$, $v = 1$, and $\mathbf{N} = \hat{\boldsymbol{v}}\hat{\boldsymbol{h}}^{\mathrm{T}}$). Depending on the relative orientation of the axes of the retarder and that of the linear analyzer, one could think of many configurations for the polarimeter. An example is illustrated by the so-called *right-handed circular analyzer*, in which the linear polarizer has its optical axis inclined at 45° to those of a quarter-wave plate ($\delta = \pi/2$). Such a device has the following Mueller matrix:

$$\mathbf{M}(\theta) = \mathbf{L}(\pi/4 + \theta)\mathbf{R}(\theta, \pi/2) = \frac{1}{2} \begin{pmatrix} 1 & 0 & 0 & 1 \\ -s_2 & 0 & 0 & -s_2 \\ c_2 & 0 & 0 & c_2 \\ 0 & 0 & 0 & 0 \end{pmatrix}, \tag{4.53}$$

where θ is again the angle between the fast axis of the retarder and the X axis. The reader can easily check that Eq. (4.53) represents an analyzer ($h = 1$) that *completely transmits* right-handed circular polarization and *completely extinguishes* left-handed circular polarization. It represents a polarizer as well: the output state

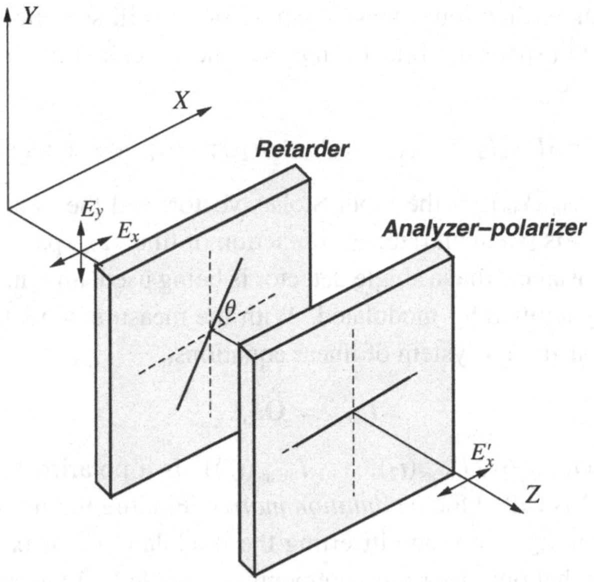

Fig. 4.5. The basic polarimeter consists of a linear retarder and a linear analyzer–polarizer.

of polarization is always $\boldsymbol{I}' = 1/2(I + V)(1, -s_2, c_2, 0)^{\mathrm{T}}$, i.e., linearly polarized light at an angle of $\theta + \pi/4$. Should the linear analyzer have been oriented at 135°, one would have obtained a *left-handed circular analyzer*.

To illustrate the importance of order in matrix multiplication, let us consider the reverse system, i.e., the linear analyzer followed by the retarder with axes at ±45°. Its Mueller matrix is

$$\mathbf{M}(\theta) = \mathbf{R}(\theta, \pi/2)\mathbf{L}(\pi/4 + \theta) = \frac{1}{2}\begin{pmatrix} 1 & -s_2 & c_2 & 0 \\ 0 & 0 & 0 & 0 \\ 0 & 0 & 0 & 0 \\ -1 & s_2 & -c_2 & 0 \end{pmatrix}. \tag{4.54}$$

Now, we have a linear analyzer ($h = 1$) whose transmittance is at a maximum for light linearly polarized at $\theta + \pi/4$, and a *left-handed* circular polarizer ($v = 1$) because it always outputs $\boldsymbol{I}' = 1/2(I - s_2 Q + c_2 U)(1, 0, 0, -1)^{\mathrm{T}}$.

4.7 Spatial and temporal modulation

A single measurement of the polarimetric signal is not enough to determine all four Stokes parameters of the incoming radiation. One has to modify the diattenuation vector of the analyzer (e.g., by modifying θ and/or δ) in order to change the polarimetric signal. At least four measurements are required. Some polarimeters, however, need more than four measurements, as we will see later. Modifications of $\boldsymbol{h}$ can be done in space, in time, or in both. The general equation for the polarimetric signal reads

$$I_{\text{meas}}(t; x) = M_{00}\left[I_{\text{in}} + h_1(t; x)Q_{\text{in}} + h_2(t; x)U_{\text{in}} + h_3(t; x)V_{\text{in}}\right], \qquad (4.55)$$

where $(I_{\text{in}}, Q_{\text{in}}, U_{\text{in}}, V_{\text{in}})^{\text{T}}$ is the input Stokes vector, and the diattenuation vector of the polarimeter is parameterized as a function of time (t), space (x), or both.

Imagine, for instance, that a single detector is being used, and that the polarimetric signal is only temporally modulated. With the measurements taken at n given times one can construct a system of linear equations,

$$\boldsymbol{I}_{\text{meas}} = \mathbf{O}\boldsymbol{I}_{\text{in}}, \qquad (4.56)$$

where $\boldsymbol{I}_{\text{meas}} \equiv (I_{\text{meas}}(t_1), I_{\text{meas}}(t_2), \ldots, I_{\text{meas}}(t_n))^{\text{T}}$ is a polarimetric signal vector and the matrix $\mathbf{O}$ is called the *modulation matrix*. Finding the input Stokes vector then involves nothing more than inverting the modulation matrix. When the polarimeter is such that only four measurements are needed, $\mathbf{O}$ has a unique inverse. In other cases, special strategies have been devised to optimize the demodulation process (see Section 5.2).

Since the basic polarimeter consists of a linear retarder and a linear analyzer, we have three basic parameters to modify with space or time, namely, the retardance of the retarder, the orientation angle of the retarder, and the orientation angle of the analyzer–polarizer.

4.7.1 Spatial modulation

In a broad sense, by spatial modulation is meant the modification within a single (two-dimensional) detector or between two (simultaneously used) detectors of some of the above free parameters. In this section, three examples are shown. Two of them can be used with a single detector, namely, the Babinet compensator and the double birefringent plate; the third example, the polarizing beam splitter, often employs two detectors.

4.7.1.1 The Babinet compensator

The Babinet compensator is a linear retarder whose retardance varies linearly in a given direction. It consists of two wedges of quartz (positive birefringence) with exactly equal acute angles that are put in contact as in Fig. 4.6, and that can be displaced with respect to each other in the x direction. The optical axes of both wedges are parallel to the input and output faces of the compensator but are orthogonal between them (one is in the plane of the paper and the other perpendicular to it). Thus, the ordinary beam for the first wedge becomes extraordinary for

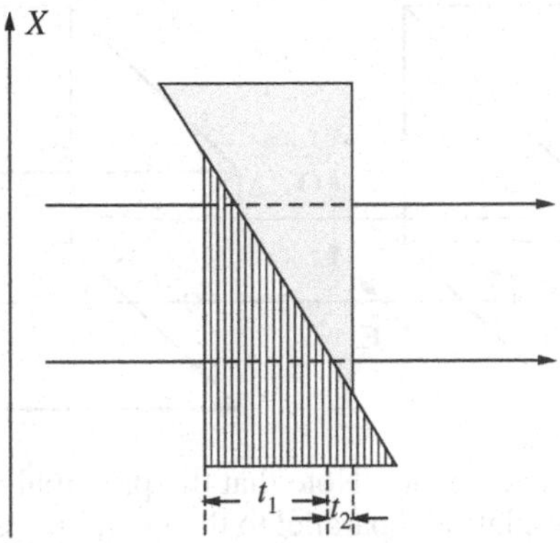

Fig. 4.6. The Babinet compensator. The optical axis of the first wedge is parallel to the vertical shadowing. The optical axis of the second is perpendicular to the plane of the figure.

the second, and vice versa. If the incoming radiation is perpendicular to the faces of the compensator, the wave normal is perpendicular to the optical axis of the medium at any time. No double refraction, then, takes place and only retardation is produced. Both the ordinary and the extraordinary rays stay in the same direction as that of the incoming radiation. Of course, if the retardance is positive for the first wedge then it is negative for the second, or vice versa. Therefore, if t_1 and t_2 are the thicknesses of the wedges at a given position x, the total retardance depends on the difference $t_1 - t_2$:

$$\delta = \frac{2\pi}{\lambda_0}\beta(t_1 - t_2) = \frac{2\pi}{\lambda_0}\beta(a + bx), \tag{4.57}$$

where a and b are constants that can easily be determined from the displacement between the wedges.

A combination of two compensators with their optical axes at 45° relative to each other plus a linear analyzer allows measurement of the four Stokes parameters. The demodulation is made in the (spatial) Fourier domain with the help of numerical filters. However, this solution is not widely used in practice because of the limited field of view for a given number of pixels of the detector (see Stenflo, 1994).

4.7.1.2 *The double birefringent plate*

If a block of quartz (or another birefringent crystal) is cut such that its optical axis forms an angle with the input and output faces of the crystal, a normally incident beam is spatially split in two at the output (see Fig. 4.7): the ordinary ray is not deviated from the original direction while the extraordinary is refracted at an angle ϕ (Sections 4.1 and 4.2). Since the optical path of the two beams is not the same, the extraordinary beam is also retarded (positively or negatively) relative to

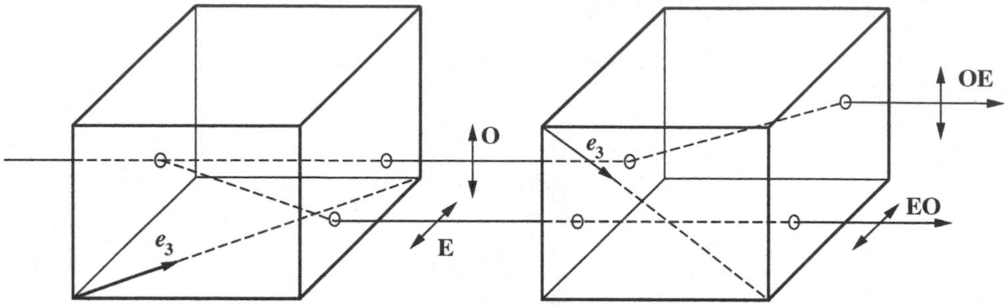

Fig. 4.7. The double birefringent plate. Note that the principal axes are parallel to the top and bottom faces for the first plate and parallel to the lateral faces for the second plate. The separation between ordinary and extraordinary rays always takes place on the principal plane.

the ordinary one depending on both β and the geometrical path of each ray. If one is interested only in spatially separating the two orthogonal states without any phase shift between them, an exactly equal plate can be placed behind the first but rotated by 90°, as in the figure. In this way, the extraordinary beam for the first plate becomes ordinary for the second and vice versa. Since the second phase shift between the two beams is the opposite to that of the first, in the end one has both beams spatially separated but no longer phase-shifted. Therefore, the double birefringent plate may play the role of a *double* linear analyzer with Mueller matrices given by

$$\mathbf{M}_{\mathrm{EO}} = \frac{1}{2}\begin{pmatrix} 1 & c_2 & s_2 & 0 \\ c_2 & c_2^2 & c_2 s_2 & 0 \\ s_2 & c_2 s_2 & s_2^2 & 0 \\ 0 & 0 & 0 & 0 \end{pmatrix} \tag{4.58}$$

for the beam which is extraordinary for the first crystal and

$$\mathbf{M}_{\mathrm{OE}} = \frac{1}{2}\begin{pmatrix} 1 & -c_2 & -s_2 & 0 \\ -c_2 & c_2^2 & c_2 s_2 & 0 \\ -s_2 & c_2 s_2 & s_2^2 & 0 \\ 0 & 0 & 0 & 0 \end{pmatrix} \tag{4.59}$$

for the beam which is ordinary for the first crystal. In Eqs (4.58) and (4.59), the notation $c_2 = \cos 2\theta$ and $s_2 = \sin 2\theta$ is kept where θ stands for the angle between the projection of the fast optical axis of the first plate on the XY plane and the positive X axis. (The Z axis is still assumed to mark the direction of the wave normal.)

The intensities of the two states of polarization can be measured simultaneously in different parts of the same detector because the output direction coincides with the input direction for both beams. However, the double image produced by the double plate implies an overlapping of the OE and EO beams coming from different points of the object. In order to avoid this effect, half of the object should be shadowed properly.

An interesting example of a practical solution is provided by the Stokes V polarimeter designed by Semel (1981), in which the double birefringent plate is preceded by a quarter-wave plate whose axes are at 45° and 135° relative (in the XY plane) to the optical axis of the first plate. Therefore, one beam experiences a right-handed circular analysis and the other a left-handed circular analysis. On output, one has a beam of intensity proportional to $(I + V)$ and another of intensity proportional to $(I - V)$ of the incoming radiation. The Semel polarimeter was designed for spectropolarimetric observations, that is, for spectroscopic observations of light whose polarization properties have been previously analyzed. For these types of observations, the polarimetric analyzer is often located just behind the spectrograph slit.† The slit is always in the focal plane of the telescope where an image of the Sun is formed, so that only light from that part of the Sun lying over the slit is allowed to enter the spectrograph. If the slit is covered by an opaque comb-like grid with holes exactly separated by the distance between the double images, overlapping in the focal plane of the spectrograph is avoided. The splitting of images should in principle be small to minimize optical aberrations in the crystal. However, an elaborate solution has been found (Semel, 1987) and the opaque grid can be as simple as one that covers half of the slit. Thus, for every observed region of the Sun we can measure $I \pm V$ at the same time in different rows of the detector.

4.7.1.3 *Polarizing beam splitter*

A similar spatial modulation can be achieved with the so-called *polarizing beam splitter*. As illustrated in Fig. 4.8, the two orthogonal, linearly polarized components of the incident beam are split in two different directions. The extraordinary beam continues in the original direction while the ordinary beam deviates by reflection at the interface between two pieces of an originally single block of birefringent crystal. The original block is cut and then stuck with an optical cement whose index of refraction is such as to produce total reflection of the ordinary beam. The two Mueller matrices each corresponding to the action of the beam splitter on each of the orthogonal components, are in fact equal to those of the double birefringent plate. Hence, a polarimeter using a beam splitter as the linear analyzer can measure two orthogonal states simultaneously but with different detectors in most cases.

† See a discussion on the location of the polarization analyzer in Section 5.1.

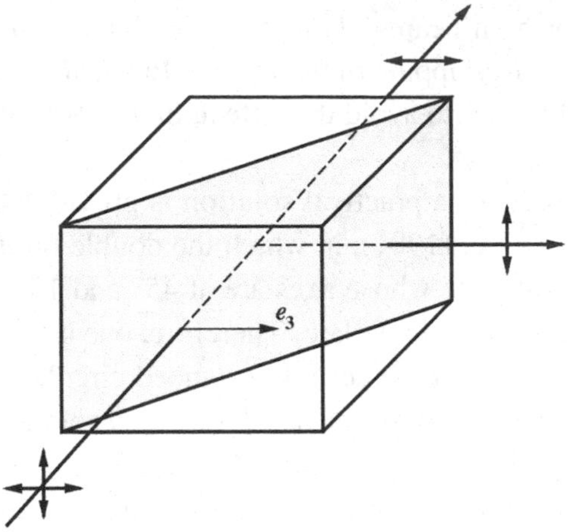

Fig. 4.8. The beam splitter. Total reflection takes place at the interface for the ordinary beam in this example.

If either a double birefringent plate or a polarizing beam splitter is used as a spatial modulator, then the four Stokes parameters cannot all be measured simultaneously. Nevertheless, one can separate intensity from polarization information by simply adding and subtracting the two output beams. To understand this, let us consider any of these systems as a double polarimeter with two Mueller matrices, $\mathbf{M}_{\mathrm{P}_1}$ and $\mathbf{M}_{\mathrm{P}_2}$. Both matrices are the products of matrices $\mathbf{M}_{\mathrm{EO}}$ and $\mathbf{M}_{\mathrm{OE}}$ of Eqs (4.58) and (4.59) with the matrix of the retarder, $\mathbf{R}$, respectively. Since the diattenuation and polarizance vectors of $\mathbf{M}_{\mathrm{OE}}$ and $\mathbf{M}_{\mathrm{EO}}$ are anti-parallel, so are the diattenuation and polarizance vectors of $\mathbf{M}_{\mathrm{P}_1}$ and $\mathbf{M}_{\mathrm{P}_2}$:

$$\mathbf{M}_{\mathrm{P}_1} = \frac{1}{2}\begin{pmatrix} 1 & \boldsymbol{h}^{\mathrm{T}} \\ \boldsymbol{h} & \boldsymbol{h}\boldsymbol{h}^{\mathrm{T}} \end{pmatrix} \begin{pmatrix} 1 & \boldsymbol{0}^{\mathrm{T}} \\ \boldsymbol{0} & \mathbf{N}_{\mathrm{R}} \end{pmatrix} = \frac{1}{2}\begin{pmatrix} 1 & \boldsymbol{h}^{\mathrm{T}}\mathbf{N}_{\mathrm{R}} \\ \boldsymbol{h} & \boldsymbol{h}\boldsymbol{h}^{\mathrm{T}}\mathbf{N}_{\mathrm{R}} \end{pmatrix}, \tag{4.60}$$

$$\mathbf{M}_{\mathrm{P}_2} = \frac{1}{2}\begin{pmatrix} 1 & -\boldsymbol{h}^{\mathrm{T}} \\ -\boldsymbol{h} & \boldsymbol{h}\boldsymbol{h}^{\mathrm{T}} \end{pmatrix} \begin{pmatrix} 1 & \boldsymbol{0}^{\mathrm{T}} \\ \boldsymbol{0} & \mathbf{N}_{\mathrm{R}} \end{pmatrix} = \frac{1}{2}\begin{pmatrix} 1 & -\boldsymbol{h}^{\mathrm{T}}\mathbf{N}_{\mathrm{R}} \\ -\boldsymbol{h} & \boldsymbol{h}\boldsymbol{h}^{\mathrm{T}}\mathbf{N}_{\mathrm{R}} \end{pmatrix}, \tag{4.61}$$

where $\boldsymbol{h} = \boldsymbol{h}_{\mathrm{EO}} = -\boldsymbol{h}_{\mathrm{OE}}$. Then, if we call $\boldsymbol{h}' \equiv \mathbf{N}_{\mathrm{R}}\boldsymbol{h}$, the polarimetric signals of the two beams are

$$I_{\mathrm{meas},1} = \frac{1}{2}(I_{\mathrm{in}} + h'_1 Q_{\mathrm{in}} + h'_2 U_{\mathrm{in}} + h'_3 V_{\mathrm{in}}), \tag{4.62}$$

$$I_{\mathrm{meas},2} = \frac{1}{2}(I_{\mathrm{in}} - h'_1 Q_{\mathrm{in}} - h'_2 U_{\mathrm{in}} - h'_3 V_{\mathrm{in}}), \tag{4.63}$$

which, after addition and subtraction give

$$I_{\mathrm{meas},1} + I_{\mathrm{meas},2} = I_{\mathrm{in}} \tag{4.64}$$

$$I_{\text{meas},1} - I_{\text{meas},2} = h'_1 Q_{\text{in}} + h'_2 U_{\text{in}} + h'_3 V_{\text{in}} \quad (4.65)$$

as we wanted to demonstrate.

In some particular cases, one may be interested in just measuring one of the three last Stokes parameters. This circumstance might be due, for example, to previous knowledge that in a given object only linearly or circularly polarized light is emitted, or that Stokes Q and Stokes U are significantly less than Stokes V and can safely be neglected. The double measurement of Eqs (4.62) and (4.63) is sufficient. As an example, the aforementioned Semel Stokes V polarimeter enters into this category. Its design is such that $h'_1 = h'_2 = 0$ and $h'_3 \neq 0$. As we shall see in Chapter 5, however, modern accurate solar polarimetry almost always requires a full polarization analysis. Therefore, the two last options for spatial modulation are not enough and are often used in combination with temporal modulation.

4.7.2 Temporal modulation

If in the general expression (4.55) for the polarimetric signal the only parameter is time, the modulation is called *temporal*:

$$I_{\text{meas}}(t) = M_{00}\left[I_{\text{in}} + h_1(t) Q_{\text{in}} + h_2(t) U_{\text{in}} + h_3(t) V_{\text{in}}\right]. \quad (4.66)$$

The components of the diattenuation vector of the polarimeter may vary with time in many different ways, of which we point out two:

- *Rotation of the retarder*. If one allows the retarder to rotate with an angular frequency Ω, the orientation angle of its fast axis varies as $\theta(t) = \Omega t$. This effects a quadratic sinusoidal modulation of the retarder Mueller matrix and hence the polarimeter Mueller matrix becomes time modulated.
- *Electro-optical modulators*. Electro-optical modulators are birefringent devices whose retardance or orientation angle vary after the application of an external electric field. Examples are piezo-elastic modulators and nematic liquid crystals (variable retardance) and ferroelectric liquid crystals (variable orientation).

Piezo-elastic modulators are made of glass in which birefringence is induced by a transducer that establishes a standing acoustic wave between the walls at the fundamental-mode frequency of the material. The oscillating pressure generates an oscillating strain, which in turn produces a variable retardance at the same frequency, typically at 50 kHz, which is kept very stable because it equals the resonant frequency of the glass block.

Liquid crystals are anisotropic molecules which can be oriented to form birefringent layers. Hence, such molecules, suitably aligned within glass plates, can be used as retardation plates. Nematic liquid crystals are made of several layers

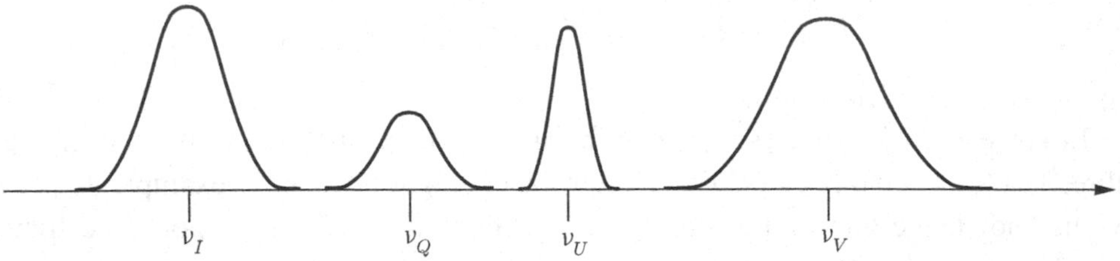

Fig. 4.9. Idealized shape of the power spectrum of a perfectly modulated polarimetric signal. The distributions for I, Q, U, and V are well separated from each other. $\nu_I = 0$ always. Scales are arbitrary.

of such molecules (typical thickness ≈ 6 μm) whose orientation is varied by the application of an external voltage. It is the orientation variation of the molecules that induces a time-varying retardance.

Ferroelectric liquid crystals are made of a very thin layer (typical thickness ≈ 2 μm) of materials that have the following properties: (1) the ferroelectric cell behaves as a constant-δ retarder; (2) once a voltage is received, the liquid crystal switches between the only two possible orientations of the axis (the only two quantum states of its spontaneous polarization vector); and (3) the flipping time between states is very fast (up to kHz).

The ideal polarimeter based on temporal modulation should be such that the three components of the diattenuation vector oscillate with frequencies that are different enough to avoid overlapping of the spectral features of I, Q, U, and V. The I spectrum is always centered at $\nu_I = 0$ (the DC component) because the first Stokes parameter is not temporally modulated [e.g., Eq. (4.66)]. This ideal situation is sketched in Fig. 4.9, where the square modulus of the Fourier transform of $I_{\text{meas}}(t)$ is drawn on the assumption that $\nu_Q < \nu_U < \nu_V$. In such a situation, simple band-pass filters allow the independent recovery of all four Stokes parameters. The separation between the amplitude and phase spectra of Stokes I, Q, U, and V depends on the polarimeter design, and, of course, some overlapping may remain. Such overlapping induces *cross-talk* between the Stokes parameters that in the end determines the polarimetric accuracy of the system.

Recommended bibliography

Born, M. and Wolf, E. (1993). *Principles of optics*, 6th edition (Pergamon Press: Oxford). Chapter 14. Sections 14.1–14.3, 10.7.3, and 10.8.

Givens, C.R. and Kostinski, A. (1993). A simple necessary and sufficient condition on physically realizable Mueller matrices. *J. Mod. Opt.* **40**, 471.

Jackson, J.D. (1975). *Classical electrodynamics*, 2nd edition (John Wiley & Sons, Inc.: New York). Chapter 7. Sections 7.3–7.5.

Landi Degl'Innocenti, E. (1992). Magnetic field measurements, in *Solar observations: techniques and interpretations*. F. Sánchez, M. Collados, and M. Vázquez (eds.) (Cambridge University Press: Cambridge). Sections 2–3.

Landi Degl'Innocenti, E. and del Toro Iniesta, J.C. (1998). Physical significance of experimental Mueller matrices. *J. Opt. Soc. Am. A* **15**, 533.

Lu, S.-Y. and Chipman, R.A. (1996). Interpretation of Mueller matrices based on polar decomposition. *J. Opt. Soc. Am. A* **13**, 1106.

Lu, S.-Y. and Chipman, R.A. (1998). Mueller matrices and the degree of polarization. *Optics Comm.* **146**, 11.

Mueller, H. (1948). The foundations of optics. *J. Opt. Soc. Am.* **38**, 661.

Semel, M. (1981). Magnetic fields observed in a sunspot and faculae using 12 lines simultaneously. *Astron. Astrophys.* **97**, 75.

Semel, M. (1987). Polarimetry and imagery through uniaxial crystals. Application to solar observations with high spatial resolution. *Astron. Astrophys.* **178**, 257.

Shurcliff, W.A. (1962). *Polarized light* (Harvard University Press & Oxford University Press: Cambridge, Mass. & London). Chapter 8, Appendix 2.

Stenflo, J.O. (1994). *Solar magnetic fields. Polarized radiation diagnostics* (Kluwer Academic Publishers: Dordrecht). Chapter 13, Sections 13.1–13.4 and 13.7.

5

Solar polarimetry

If it were not for its magnetic field, the Sun would be as dull a star as most astronomers think it is.

—R. Leighton, 1965.

Polarimetric accuracy is one of the most important goals of modern astronomy. The definition itself of polarimetric accuracy, however, is difficult since we mostly measure polarization differences and are uncertain in establishing the zero level, which is often set by convention. Hence, by "accuracy" we shall understand the sensitivity to variations of the polarization level. Besides the greatest polarimetric accuracy, every astronomical observation should ideally pursue the highest spectral, spatial, and temporal resolution with the widest spatial and spectral coverage. However, all these goals are hard to accomplish at the same time and one always needs to compromise depending on the specific objectives a given observation is aimed at. The amount of available photons from the Sun is never sufficient. In fact, it is equal per resolution element to that from a scarcely resolved star of the same effective temperature. This observational fact is easy to understand (e.g., Mihalas, 1978) if one takes into account the invariance with distance of the specific intensity (energy per unit normal surface, per unit time, per unit frequency interval, per unit solid angle) and its proportionality to the photon distribution function (number of photons per unit volume, per unit frequency interval, per unit solid angle).

Solar polarimetry is, of course, a part of the game and has several limiting factors that govern the final accuracy of the measurements. These factors can be categorized into two main groups, namely, the *environmental polarization* and the characteristics of the polarization analysis system itself. Both groups are discussed separately in this chapter. A few words on wavelength dependence are of common interest to both groups, however.

Most optical elements found by light on its path from the source to the detector, including the Earth's atmosphere, have polarization properties dependent on wavelength. For instance, from the definition of retardance (4.7), it is easy to see that a quarter-wave plate for a given wavelength could not be such for other wavelengths even if the refractive indices were not wavelength dependent. But n does depend on wavelength, as has been well known since Newton's experiments on prismatic colors. This further complicates the design of polarimeters if one wants to keep them achromatic, or implies differences in those designs that are contingent on the working spectral range (e.g., visible or infrared) of the polarimeter. The details concerning chromaticity are beyond the scope of this chapter but should be borne in mind by the reader. A discussion about absorption and dispersion, and hence on the wavelength dependence of refractive indices, is presented in Chapter 6. In most practical cases, calibration of both the environment and the analyzer is needed prior to properly interpreting the results at given (different) wavelengths.

5.1 Environmental polarization

The environment influences astronomical polarimetry and even may jeopardize the reliability of the results. By "environment" we mean all the systems other than the polarimeter that may alter the polarization state of the incoming light. This is the opposite situation to that of laboratory polarimetry, where all the conditions external to the experiment can be controlled. Therefore, astronomical polarimetry should not only take care of the analyzer Mueller matrix but also of the Mueller matrices of all other systems involved in the optical train of observations. Within environmental systems we can clearly distinguish the Earth's atmosphere and the instrumental set-up. Consequently we can speak of atmospheric polarization, also called *seeing-induced* polarization, and of *instrumental* polarization. Seeing-induced polarization can be avoided only if the observations are carried out on board a spacecraft or a stratospheric balloon, i.e., outside the disturbing effects of the Earth's atmosphere. Instrumental polarization can be avoided only if the polarimetric analysis is performed before the light encounters any feed or beam-steering optics, but this is unfeasible because of the usually small apertures of polarimeters. Minimizing both seeing-induced and instrumental polarization is more likely, however.

A typical observation is affected by the Earth's atmosphere, the telescope, the polarimeter, the spectrograph, and the detector. If the Mueller matrices of these elements are respectively called $\mathbf{M}_\mathrm{A}$, $\mathbf{M}_\mathrm{T}$, $\mathbf{M}_\mathrm{P}$, $\mathbf{M}_\mathrm{S}$, and $\mathbf{M}_\mathrm{D}$, and light streams through these systems in that order, the ensemble Mueller matrix is then

$$\mathbf{M} = \mathbf{M}_\mathrm{D}\mathbf{M}_\mathrm{S}\mathbf{M}_\mathrm{P}\mathbf{M}_\mathrm{T}\mathbf{M}_\mathrm{A}. \tag{5.1}$$

The solar Stokes parameters, $\boldsymbol{I}_{\rm sun}$, are thus related to the observed Stokes parameters, $\boldsymbol{I}_{\rm obs}$, by $\boldsymbol{I}_{\rm sun} = \mathbf{M}^{-1}\boldsymbol{I}_{\rm obs}$. The accuracy of the measurements then depends on an accurate knowledge of matrix $\mathbf{M}$ (and on the existence of its inverse matrix!).

It is easily seen that if one (or some) of the systems is (are) not properly taken into account (or even neglected) the final result may differ from reality. Nevertheless, a distinction must be made among those systems located in the optical train before and after the polarimeter. By far, the most relevant degrading effects arise from the systems located before the polarimeter, as we shall see. After interaction with those systems prior in the optical train to the analyzer, the Stokes parameters become linear combinations of the original ones [Eq. (4.29)]. The polarimeter thus measures "falsified" Stokes parameters. We must know exactly how $\boldsymbol{I}_{\rm sun}$ has been modified in order to correct for this modification. However, on output, the analyzer *encodes* the polarization information in a known way. Typically, the output state of polarization from the polarimeter is always the same and is the intensity of light which carries the information on the object's polarization. Hence, only the gain or transmittance properties of the subsequent systems in the optical train need to be taken into account for a few given states of polarization. Therefore, locating the polarimeter in front of the telescope would seem to be good advice. [Note that the order of matrices in Eq. (5.1) would necessarily have to be altered.] Unfortunately, such a solution is not feasible in practice since polarimeter apertures are usually much smaller than telescope apertures. Polarimeters are therefore located as early in the optical train as possible in order to minimize instrumental effects.

The environmental polarization is not only relevant to polarimetric observations. Imagine carrying out standard spectroscopic observations without performing any polarization analysis. If light coming from the object were partially polarized, the observed intensity spectrum would indeed be a linear combination of the object Stokes parameters [e.g., Eq. (4.66)] and not just the intensity, as we would like it to be. The problem becomes even more significant if we take into account that the degree of polarization generally varies across the profile of a single spectral line (see an excellent discussion of the problem in Sánchez Almeida and Martínez Pillet, 1992). Broad-band (i.e., photometric) observations may *sensu stricto* suffer from the same problem: differences in the measured intensities of two regions of the Sun may come from differences in the degree of polarization. Nonetheless solar broad-band polarization is not as large as spectral line polarization, except perhaps in such features as sunspots.

5.1.1 Seeing-induced polarization

Sunlight is scattered and refracted by terrestrial atmospheric particles. In general, scattering processes induce polarization effects, so $\mathbf{M}_{\rm A}$ might be non-diagonal.

Note that a Mueller matrix proportional to the identity matrix represents only a scaling of the Stokes vector. Pure absorption processes have such Mueller matrices. Fortunately, the diagonal matrix elements of $\mathbf{M}_A$ are almost identical and the non-diagonal ones can safely be ignored for solar observations (see Martínez Pillet, 1992): atmospheric scattering is mostly single and, for the angles of interest (the Sun subtends an angular diameter of approximately half a degree), non-diagonal terms are of the order of 10^{-5} or less than diagonal ones. The contributions from refractive index perturbations are even less important. Therefore, for the purposes of solar polarimetry up to the 10^{-5} level of accuracy, a *frozen* atmosphere has $\mathbf{M}_A \propto \mathbb{1}$.

But the Earth's atmosphere is not static. Temporal fluctuations produce wavefront distortions which in turn induce spatial image smearing and spurious polarization features. If the solar atmosphere is assumed to be static for the time interval of measurements – which is a reasonable assumption – spatial smearing can be characterized by averages ($\langle \delta I_{\text{seeing}} \rangle$, $\langle \delta Q_{\text{seeing}} \rangle$, $\langle \delta U_{\text{seeing}} \rangle$, and $\langle \delta V_{\text{seeing}} \rangle$) that represent the contributions to every resolution element from its surroundings. In other words, the Stokes parameters measured on every pixel are those intrinsic to the solar region plus these spatial averages. Since the polarization state is not spatially constant over the solar surface, spatial resolution is lost not only in intensity images but also in all four Stokes parameter images. Besides this spatial smearing, spurious polarization features may appear in the observations as a consequence of slow modulation during the measurement process. Let us characterize this effect by a contribution $\delta \boldsymbol{I}_{\text{seeing}}(t)$ to the observations. If we then call $\boldsymbol{I}_A$ the Stokes vector after the light has traveled through the atmosphere, we have

$$\boldsymbol{I}_A = \boldsymbol{I}_{\text{sun}} + \langle \delta \boldsymbol{I}_{\text{seeing}} \rangle + \delta \boldsymbol{I}_{\text{seeing}}(t), \tag{5.2}$$

where we have neglected possible scaling factors.

Spatial smearing can be mitigated with high spatial resolution techniques such as adaptive optics, or speckle interferometry, formerly developed for unpolarized observations. We shall hereafter neglect the contribution of $\langle \delta \boldsymbol{I}_{\text{seeing}} \rangle$ in Eq. (5.2) on the assumption that corrections for image smearing have already been made. These high resolution techniques, however, are not able to avoid intermixing among Stokes parameters that produces spurious polarization features. The fluctuating polarization, $\delta \boldsymbol{I}_{\text{seeing}}(t)$, induced by seeing contributes with frequencies of the order of 100 Hz to the signal thus broadening the spectral content of all four Stokes parameters.† Fast temporal modulation, then, is required at frequencies higher than

† Although the terms are misleadingly similar, the reader must distinguish between *spectrum* referring to the Fourier transform of the polarimetric signal as a function of time and *spectrum* referring to the wavelength dependence of the Stokes parameters.

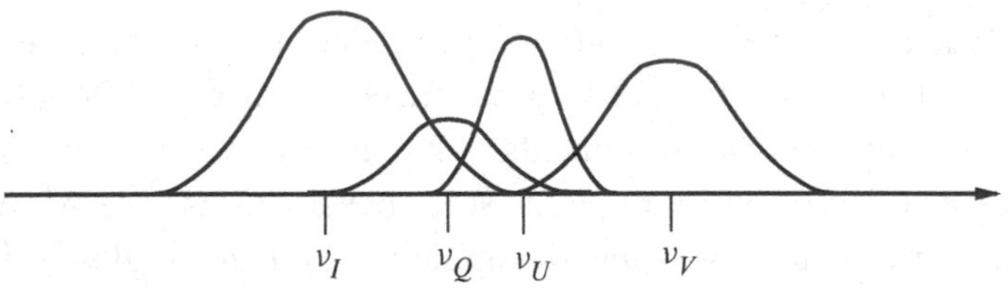

Fig. 5.1. Idealized view of the influence of seeing on the amplitude spectrum of the polarimetric signal. The individual distributions are broadened and overlap (cross-talk) with each other. $\nu_I = 0$ always. Scales are arbitrary.

that of the seeing to avoid the problem effectively. If the modulation frequency is not high enough, residual cross-talk remains, as illustrated in Fig. 5.1: the spectral content of I, Q, U, and V overlap and the demodulation process is unable to get rid of contamination from other Stokes parameters coming from a (given) single resolution element. Although Fig. 5.1 shows only the idealized amplitude spectrum, the phase spectrum must also be taken into account. In fact, it may help in practice to disentangle the information corresponding to the different Stokes parameters that overlap in the amplitude spectrum.

To get a better grasp of the problem in the measurement, rather than in the frequency, domain, consider the following example.‡ Imagine that we are simply measuring Stokes $I+V$ and $I-V$ at two different times, and that spatial smearing has already been corrected for. That is, our ideal polarimeter (a linear retarder plus a linear analyzer with axis at 45° or 135° relative to the axes of the retarder) has diattenuation vectors $\boldsymbol{h}(t_1) = (0, 0, 1)^{\mathrm{T}}$ and $\boldsymbol{h}(t_2) = (0, 0, -1)^{\mathrm{T}}$. Equation (4.55) then gives

$$\begin{aligned} I_{\rm meas}(t_1) &= \frac{k}{2}\left[(I_{\rm sun} + \delta I_{\rm seeing,1}) + (V_{\rm sun} + \delta V_{\rm seeing,1})\right], \\ I_{\rm meas}(t_2) &= \frac{k}{2}\left[(I_{\rm sun} + \delta I_{\rm seeing,2}) - (V_{\rm sun} + \delta V_{\rm seeing,2})\right], \end{aligned} \tag{5.3}$$

where k stands for the gain of the detector, $\delta I_{\rm seeing,\it i} \equiv \delta I_{\rm seeing}(t_i)$, and $\delta V_{\rm seeing,\it i} \equiv \delta V_{\rm seeing}(t_i)$. Addition and subtraction of Eqs (5.3) yields the observed Stokes parameters

$$\begin{aligned} I_{\rm obs} &= kI_{\rm sun} + \frac{k}{2}(\delta I_{\rm seeing} - \delta V'_{\rm seeing}), \\ V_{\rm obs} &= kV_{\rm sun} + \frac{k}{2}(\delta V_{\rm seeing} - \delta I'_{\rm seeing}), \end{aligned} \tag{5.4}$$

‡ Based on Collados (1999).

where

$$\delta I_{\text{seeing}} \equiv \delta I_{\text{seeing},1} + \delta I_{\text{seeing},2},$$

$$\delta V_{\text{seeing}} \equiv \delta V_{\text{seeing},1} + \delta V_{\text{seeing},2},$$

$$\delta I'_{\text{seeing}} \equiv \delta I_{\text{seeing},2} - \delta I_{\text{seeing},1},$$

$$\delta V'_{\text{seeing}} \equiv \delta V_{\text{seeing},2} - \delta V_{\text{seeing},1}.$$

Equations (5.4) show how $V \longrightarrow I$ and $I \longrightarrow V$ cross-talk appears. If a typical value for the errors introduced by seeing are of the order $\delta I_{\text{seeing},i}/I_{\text{sun}} = \delta V_{\text{seeing},i}/V_{\text{sun}} \approx 10^{-2}$, the ratios between the two terms on the right-hand side of Eqs (5.4) are

$$\begin{aligned} \frac{k(\delta I_{\text{seeing}} - \delta V'_{\text{seeing}})}{2kI_{\text{sun}}} &\approx 10^{-2}, \\ \frac{k(\delta V_{\text{seeing}} - \delta I'_{\text{seeing}})}{2kV_{\text{sun}}} &\approx 10^{-1}, \end{aligned} \tag{5.5}$$

since $V_{\text{sun}}/I_{\text{sun}}$ is of the order 10^{-1} for most solar cases. Therefore, an unacceptable $I \longrightarrow V$ cross-talk, an order of magnitude greater than $\delta I_{\text{seeing}}/I_{\text{sun}}$, is obtained.

Spatial modulation helps in getting rid of $I \longrightarrow V$ cross-talk. To illustrate this, imagine that our Stokes V polarimeter now has a beam splitter or a double birefringent plate as a linear analyzer. In such a case, $I \pm V$ can be measured at the same time. According to the discussion in Section 4.7.1, at time t_1 we have two diattenuation vectors, $\boldsymbol{h}_1(t_1) = (0, 0, 1)^{\text{T}}$ and $\boldsymbol{h}_2(t_1) = (0, 0, -1)^{\text{T}}$, so that

$$\begin{aligned} I_{\text{obs}}(t_1) &= I_{\text{meas},1}(t_1) + I_{\text{meas},2}(t_1) = k(I_{\text{sun}} + \delta I_{\text{seeing},1}), \\ V_{\text{obs}}(t_1) &= I_{\text{meas},1}(t_1) - I_{\text{meas},2}(t_1) = k(V_{\text{sun}} + \delta V_{\text{seeing},1}). \end{aligned} \tag{5.6}$$

Note that the cross-talk has now clearly disappeared.

So far we have been discussing $I \longrightarrow V$ contamination. Should we have considered measurements of $I \pm Q$ or $I \pm U$, similar conclusions would have been reached. $I \longrightarrow Q, U, V$ cross-talk is the most relevant since $I > Q, U, V$. More elaborate calculations show that temporal modulation at video frame rates together with spatial modulation with a beam splitter can keep $V \longrightarrow (Q, U) < 10^{-3} I$ and $(Q, U) \longrightarrow V$ and $(Q, U) \longrightarrow (Q, U)$ at even lower levels (Lites, 1987). In summary, correction of seeing-induced polarization can be effected either with very fast temporal modulation or with spatio–temporal modulation. Either way is chosen according to the polarimeter design, depending on further requirements of the observations such as spatial and/or spectral resolution.

5.1.2 Instrumental polarization

Let us discuss now the important polarization effects produced by the telescope, or, more generally, the image-forming system.

5.1.2.1 Telescope polarization

Telescope seeing (air turbulence) within the telescope tube can introduce similar effects to those produced by the Earth's atmosphere. On the assumption that these effects are corrected for in the same way as atmospheric seeing, or, better still, that turbulence is avoided by keeping the telescope tube under vacuum or by some other means such as an open-air tube, the most important degrading effects introduced by the image-forming system are due to oblique reflections on metallic surfaces (i.e., mirrors) and imperfections and stresses in glass windows.

Metallic surfaces act both as partial polarizers (Section 4.6.3) and as retarders (Section 4.6.4). The general Mueller matrix for reflection by a metallic mirror can be found, for example, in Stenflo (1994). Different rays of the same collimated beam reaching different parts of the mirror may suffer different modifications in the state of polarization of the beam. Nonetheless, those different rays contribute *coherently*† to the final polarization signal. It can then be shown that revolution symmetry makes the various contributions cancel out (Sánchez Almeida and Martínez Pillet, 1992). The best solution is thus to use "polarization-free" telescopes such as the Franco–Italian THEMIS,‡ in which the polarization analysis is carried out just after reflection on revolution-symmetry mirrors. Other solar telescopes have not been specifically designed for avoiding instrumental polarization. Although most main and secondary mirrors have revolution symmetry, many have flat mirrors such as coelostats or other beam-folding optics and even off-axis main mirrors through the light path.

Windows are usually made of glass, i.e., a dielectric. For small incident angles, the Mueller matrix for the transmitted radiation is diagonal and is thus of no concern for the polarimetric analysis. However, mechanical tensions and stresses may induce inhomogeneous birefringence, which makes such windows behave on average as retarders.

Since oblique metallic surfaces, windows, and other optical imperfections are present in polarization-free telescopes, calibration of the polarization properties of the whole image-forming system is mandatory in all cases. One usually proceeds by modeling the Mueller matrix $\mathbf{M}_\mathrm{T}$ of the system. This theoretical model depends on several free parameters that are then fitted with specially designed observations that may need some additional polarizing optics. Note that among the free

† Coherent superposition of rays cannot be treated with the Mueller matrix formalism; the Jones matrix formalism is used instead. See the comment in Section 3.4.

‡ Information on the telescope can be found in Mein and Rayrole (1985).

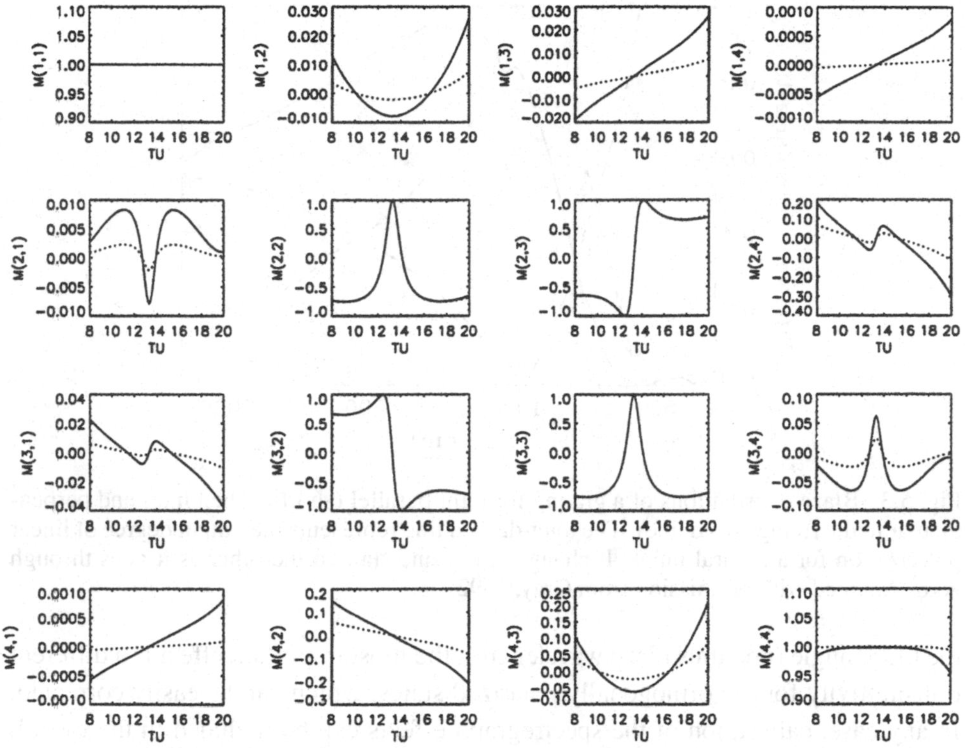

Fig. 5.2. Daily variation of $\mathbf{M}_\mathrm{T}$ matrix elements for the VTT telescope on 1 July 1988 at two wavelengths: 500 nm (solid line) and 1.56 μm (dotted line). Courtesy of M. Collados.

parameters one may find the date and time of observations. This is so because of the varying orientation angles that the different optical elements may have in order to point to the Sun. As an example, Fig. 5.2 shows the measured Mueller matrix of the German Vacuum Tower Telescope (VTT)† through the day for a given date and for two wavelengths, namely, 500 nm (solid lines) and 1.56 μm (dotted lines). Each panel corresponds to one of the matrix elements of $\mathbf{M}_\mathrm{T}$.

5.1.2.2 Spectrograph polarization

The polarization properties of gratings may be of importance for some observations, although for time-modulated signals they are not very much of a problem. Gratings have different blaze distributions and transmittivities for light parallel or perpendicular to the ruling (see Fig. 5.3). Moreover, they act as partial linear polarizers whose properties depend strongly on wavelength. For typical solar observations, in which the wavelength span is usually small and incidence is very close to

† Information on this telescope may be found in Schröter *et al.* (1985).

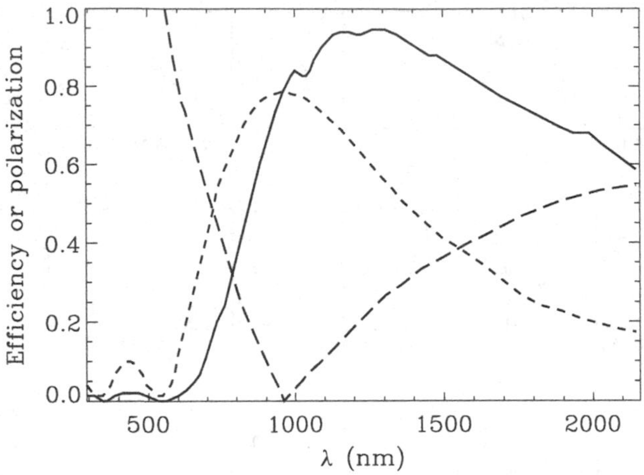

Fig. 5.3. Blaze distributions of a grating for light parallel (short-dashed line) and perpendicular to the ruling (solid line). The long-dashed line represents the output degree of linear polarization for a natural input. It changes from one state to the other as it goes through zero. Adapted with permission from Gray, 1992.

the blaze angle (and thus blaze wavelength), the most important effect is a different transmittivity for the orthogonally polarized states, which can be easily corrected. In any case, calibration of the spectrograph effects can be included in the overall calibration described above for the image-forming system.

5.1.2.3 Detector polarization

Apart from limiting the photometric accuracy, detectors can induce significant polarimetric problems if all four Stokes parameters are not measured on the same resolution element. Two-dimensional quantum detectors such as charge coupled devices (CCDs) require gain-table corrections in order to account for the different sensitivities of the individual pixels. Flat-fielding accuracies are often of the order of 10^{-2}. Therefore, the degree of polarization suffers from exceedingly large gain-table uncertainties. To better understand this effect, let us consider a similar example to that described in Section 5.1.1.‡ Imagine that we are measuring $I + V$ and $I - V$ with two detectors or on different parts of a single detector. The two measurements will be affected by gain factors k that differ by δk, the flat-field accuracy:

$$\begin{aligned} I_{\text{meas},1} &= \frac{k}{2}(I_{\text{sun}} + \delta I_{\text{seeing}} + V_{\text{sun}} + \delta V_{\text{seeing}}), \\ I_{\text{meas},2} &= \frac{(k + \delta k)}{2}(I_{\text{sun}} + \delta I_{\text{seeing}} - V_{\text{sun}} - \delta V_{\text{seeing}}). \end{aligned} \quad (5.7)$$

‡ This example is also taken from Collados (1999).

Addition and subtraction of $I_{\mathrm{meas},1}$ and $I_{\mathrm{meas},2}$ give

$$\begin{aligned} I_{\mathrm{obs}} &= \left(k+\frac{\delta k}{2}\right)(I_{\mathrm{sun}}+\delta I_{\mathrm{seeing}}) - \frac{\delta k}{2}(V_{\mathrm{sun}}+\delta V_{\mathrm{seeing}}), \\ V_{\mathrm{obs}} &= \left(k+\frac{\delta k}{2}\right)(V_{\mathrm{sun}}+\delta V_{\mathrm{seeing}}) - \frac{\delta k}{2}(I_{\mathrm{sun}}+\delta I_{\mathrm{seeing}}). \end{aligned} \tag{5.8}$$

The important cross-talk components are in the second terms on the right-hand side of the equations. Again, since $V_{\mathrm{sun}}/I_{\mathrm{sun}} \approx 10^{-1}$, the $V \longrightarrow I$ contamination is of the acceptable order of 10^{-2} of $\delta k/k$, but the $I \longrightarrow V$ contamination is considerable and must be avoided or, at least, mitigated. Spatial modulation alleviates seeing-induced cross-talk but introduces gain-table uncertainties.

A possible solution that is used in practice (Elmore *et al.*, 1992; Martínez Pillet *et al.*, 1999) comes from a mixed scheme in which spatial and temporal modulation are performed. The first is employed in order to minimize seeing-induced cross-talk and the second to minimize gain-table uncertainties. A very good and didactic example of the usefulness of this mixed modulation scheme is provided by the method proposed by Semel *et al.* (1993; see also Donati *et al.*, 1990 and Bianda *et al.*, 1998), in which the two beams leaving a double birefringent plate or a beam splitter are interchanged. Following our example, we would now have four measurements similar in pairs to those of Eqs (5.7). The only difference is that from the first to the second pair, the role of both detectors (or both parts of the single detector) are interchanged. Therefore, accounting now for seeing-induced effects,

$$\begin{aligned} I_{\mathrm{meas},1}(t_1) &= \frac{k}{2}\left[(I_{\mathrm{sun}}+\delta I_{\mathrm{seeing},1}) + (V_{\mathrm{sun}}+\delta V_{\mathrm{seeing},1})\right], \\ I_{\mathrm{meas},2}(t_1) &= \frac{(k+\delta k)}{2}\left[(I_{\mathrm{sun}}+\delta I_{\mathrm{seeing},1}) - (V_{\mathrm{sun}}+\delta V_{\mathrm{seeing},1})\right], \\ I_{\mathrm{meas},1}(t_2) &= \frac{k}{2}\left[(I_{\mathrm{sun}}+\delta I_{\mathrm{seeing},2}) - (V_{\mathrm{sun}}+\delta V_{\mathrm{seeing},2})\right], \\ I_{\mathrm{meas},2}(t_2) &= \frac{(k+\delta k)}{2}\left[(I_{\mathrm{sun}}+\delta I_{\mathrm{seeing},2}) + (V_{\mathrm{sun}}+\delta V_{\mathrm{seeing},2})\right]. \end{aligned} \tag{5.9}$$

Adding and subtracting the first and the third measurements on the one hand, the second and the fourth measurements on the other, and averaging the results, one finally obtains

$$\begin{aligned} I_{\rm obs} &= (k+\delta k)\left(I_{\rm sun}+\frac{\delta I_{\rm seeing}}{2}\right)+\frac{\delta k\delta V_{\rm seeing}}{4}, \\ V_{\rm obs} &= (k+\delta k)\left(V_{\rm sun}+\frac{\delta V_{\rm seeing}}{2}\right)+\frac{\delta k\delta I_{\rm seeing}}{4}, \end{aligned} \tag{5.10}$$

where we have adopted the same conventions for notation as in Section 5.1.1. The second terms on the right-hand sides of Eqs (5.10) are three and two orders of magnitude less than gain-table uncertainties, respectively, and can be neglected. Therefore, the degree of circular polarization turns out to be independent of the flat-field, and cross-talk between Stokes I and Stokes V disappears. The errors are only introduced by the residual perturbations arising from seeing if time modulation is not fast enough. A further increase in the signal-to-noise ratio (S/N) of measurements can be achieved simply by using longer individual exposures or by repeating them in cycles so that on-line averaging can be performed of those measurements corresponding to the same state of the analyzer.

5.2 The polarization analysis system

The heart of polarimetric analysis is of course the polarimeter itself. If all environmental polarization is corrected for then the final accuracy and S/N of the results is based critically on polarimeter design. Depending on the available devices and on the target of the observations, the Stokes analyzer can be devised so as to measure two, three, or all four Stokes parameters. Nevertheless, we have come to appreciate in this chapter that full Stokes polarimetry is necessary for properly correcting environmental effects. Hence, we hereafter assume that we are dealing with polarimeters that measure all four Stokes parameters.

Let us go back to the fundamental equation (4.56) of polarization analysis.† This equation may be termed the modulation equation and the set of n measurements, $\boldsymbol{I}_{\rm meas}$, the modulation scheme of the system. According to Eq. (4.55), the modulation matrix, $\mathbf{O}$, is made up of the first rows of the analyzer Mueller matrix for each of the measurement steps, namely,

$$\mathbf{O}=\begin{pmatrix} 1 & \boldsymbol{h}_1^{\rm T} \\ 1 & \boldsymbol{h}_2^{\rm T} \\ \vdots & \vdots \\ 1 & \boldsymbol{h}_n^{\rm T} \end{pmatrix}, \tag{5.11}$$

where we have assumed $M_{00}=1$, since it acts only as a scaling factor.

† The remaining part of this chapter is based on a paper by del Toro Iniesta and Collados (2000). I refer the interested reader to that paper for the relevant references.

As pointed out in Section 4.7, the problem is then reduced to inverting the modulation matrix. If $n = 4$, **O** has a unique inverse, **D**, the demodulation matrix. The existence of **D** is ensured, for the modulation matrix is necessarily non-singular. The linear equation system (4.56) is assumed to have a solution: all four input Stokes parameters are measured. If instead $n > 4$ then an infinite number of matrices **D** exists for which $\mathbf{DO} = \mathbb{1}$ [i.e., there is an infinite number of solutions for Eq. (4.56)]. At this point, the designer has to make a decision on which modulation scheme is more suited to the purposes of the polarimeter. That is, the number, type, and extent of the individual exposures need to be chosen according to the requirements for S/N and polarimetric accuracy which in turn depend on other requirements such as temporal, spatial, and spectral resolution. We shall see that optimization of matrices **O** and **D** is possible theoretically. This optimization provides useful hints on practical design.

The S/N can be increased by averaging a number of independent determinations of $\boldsymbol{I}_{\mathrm{in}}$.† Let us concentrate, then, on polarimetric accuracy and look for modulation schemes that minimize the uncertainties, σ_i, in the determined Stokes parameters ($i = 1$ for I_{in}, $i = 2$ for Q_{in}, $i = 3$ for U_{in}, and $i = 4$ for V_{in}), due to the ever-present intrinsic noise of the measurements.

Inversion of Eq. (4.56) provides

$$\boldsymbol{I}_{\mathrm{in}} = \mathbf{D}\boldsymbol{I}_{\mathrm{meas}}. \tag{5.12}$$

The uncertainties, σ_i, are then given by

$$\sigma_i^2 = \sigma^2 \sum_{j=1}^{n} D_{ij}^2, \tag{5.13}$$

where we have assumed that every $I_{\mathrm{meas},j}$, $j = 1, 2, \ldots, n$, has the same uncertainty, σ, owing, for instance, to photon noise. The uncertainty in every Stokes parameter is then given by the sum of squares of the corresponding row of the demodulation matrix. The smaller σ_i, the better.

Equation (5.13) does not allow direct comparison between modulation schemes using different numbers of measurements. One would expect that the larger n, the smaller σ_i, but this is not clearly seen in the equation referred to. It is thus convenient to define an *efficiency of the modulation scheme* as the four-vector $\boldsymbol{\xi}$ of components

$$\xi_i = \left(n \sum_{j=1}^{n} D_{ij}^2 \right)^{-1/2}. \tag{5.14}$$

† Note that we are using the suffix "in" instead of "sun" as before. Hence, we are implicitly admitting that corrections for environmental polarization are not perfect.

On average, the contribution of every individual measurement to the final uncertainties should be $\bar{\sigma}_i^2 = n\sigma_i^2$. Equations (5.13) and (5.14) then give

$$\bar{\sigma}_i^2 = \frac{\sigma^2}{\xi_i^2}. \tag{5.15}$$

Equation (5.15) clearly indicates that two modulation schemes having the same polarimetric efficiencies do indeed provide equal individual contributions to the Stokes parameter uncertainties. That with the higher number of measurements presents smaller uncertainties in the Stokes parameters. Optimization of the modulation scheme thus requires maximization of polarimetric efficiencies.

5.2.1 The optimum modulation matrix

The optimum modulation scheme has a modulation matrix, **O**, such that

$$\mathbf{A} \equiv \mathbf{O}^{\mathrm{T}}\mathbf{O} = n \begin{pmatrix} \xi_{\max,1}^2 & 0 & 0 & 0 \\ 0 & \xi_{\max,2}^2 & 0 & 0 \\ 0 & 0 & \xi_{\max,3}^2 & 0 \\ 0 & 0 & 0 & \xi_{\max,4}^2 \end{pmatrix}, \tag{5.16}$$

where the maximum efficiencies, $\xi_{\max,i}$, are given by

$$\xi_{\max,i}^2 = \frac{\sum_{j=1}^{n} O_{ji}^2}{n}. \tag{5.17}$$

Moreover, it becomes clear that if the analyzer is *ideal*, that is, if $h_j = 1, \ \forall j = 1, 2, \ldots n$ (see Section 4.6.2) then Eq. (5.17) leads to

$$\begin{aligned} \xi_{\max,1} &= 1, \\ \sum_{i=2}^{4} \xi_{\max,i}^2 &= 1. \end{aligned} \tag{5.18}$$

To demonstrate the result expressed by Eqs (5.16) and (5.17) let us maximize the polarimetric efficiencies by minimizing the quantities

$$d_i^2 \equiv \sum_{j=1}^{n} D_{ij}^2. \tag{5.19}$$

Since d_i^2 has no non-trivial minimum, let us search for minima subject to the condition that the diagonal elements of **DO** are unity. In other words, let us seek the minimum of the function

$$f_i \equiv d_i^2 - x \left(\sum_{j=1}^{n} D_{ij} O_{ji} - 1 \right), \tag{5.20}$$

where x is a Lagrange multiplier. Taking derivatives with respect to D_{ij}, requiring the term in parentheses to be zero, and remembering that the second derivatives are always positive and equal to 2, the minimum of d_i (under that binding condition) is found if

$$D_{ij} = \frac{O_{ji}}{\sum_{j=1}^{n} O_{ji}^2}, \tag{5.21}$$

whence

$$d_{\min,i}^2 = \frac{1}{\sum_{j=1}^{n} O_{ji}^2}, \tag{5.22}$$

which directly yields Eq. (5.17). But Eq. (5.21) does not give a proper demodulation matrix unless $\sum_{j=1}^{n} D_{ij} O_{jk} = \delta_{ik}$, where δ_{ik} is the Kronecker delta. This last condition is fulfilled if and only if $\sum_{j=1}^{n} O_{ji} O_{jk} = A_{ik} = A_{ii}\delta_{ik}$, i.e., if and only if Eq. (5.16) holds.

Of course, not every modulation scheme can be optimal. An example of such an optimal scheme is that with a modulation matrix having its last three columns with elements of one and only one magnitude, namely, q, u, and v, and the two possibilities for sign appear at least twice (only twice if $n = 4$). Also, every appearance must be accompanied (in a given row) by one of the signed values of the other two. For instance, when $n = 4$,

$$\mathbf{O} = \begin{pmatrix} 1 & q & u & v \\ 1 & q & -u & -v \\ 1 & -q & u & -v \\ 1 & -q & -u & v \end{pmatrix}, \tag{5.23}$$

or any equivalent matrix. For this modulation matrix, matrix $\mathbf{A}$ becomes

$$\mathbf{A} = n \ \mathrm{diag}(1, q^2, u^2, v^2). \tag{5.24}$$

As one may be interested in Q, U, and V being obtained with the same efficiency, q, u, and v should be all equal to $1/\sqrt{3}$ since, according to the second of Eqs (5.18), $q^2 + u^2 + v^2 = 1$.

Another example fulfilling Eqs (5.16) and (5.17) is that of a modulation scheme with which Q, U, and V are measured independently, that is, a scheme for which the modulation matrix reads

$$\mathbf{O} = \begin{pmatrix} 1 & 1 & 0 & 0 \\ 1 & -1 & 0 & 0 \\ 1 & 0 & 1 & 0 \\ 1 & 0 & -1 & 0 \\ 1 & 0 & 0 & 1 \\ 1 & 0 & 0 & -1 \end{pmatrix}. \tag{5.25}$$

In this case,

$$\mathbf{A} = 6 \ \mathrm{diag}(1, 1/3, 1/3, 1/3). \tag{5.26}$$

5.2.2 The optimum demodulation matrix

Optimum modulation schemes may not be attainable in practice. On the one hand, most of the block elements of the analyzer may not behave ideally; on the other, measurements generally require the finite integration of non-linear functions. In both cases, the diattenuation vectors $\boldsymbol{h}_j, \ \ j = 1, 2, \ldots n$ are no longer unit vectors. The relevant problem is now the following. Let us consider a given (non-ideal) modulation matrix, **O**, and let us look for the best demodulation matrix, **D**, that maximizes the efficiencies.

Following the philosophy of Section 5.2.1, we shall now search for the minimum of

$$f_i \equiv d_i^2 - \sum_{k=1}^{4} x_{ik} \left(\sum_{j=1}^{n} D_{ij} O_{jk} - \delta_{ik} \right), \tag{5.27}$$

where, again, x_{ik} are Lagrange multipliers. Carrying out similar calculations as for Eq. (5.20), one finds that the optimum demodulation matrix is given by

$$\mathbf{D}_{\mathrm{opt}} = (\mathbf{O}^{\mathrm{T}}\mathbf{O})^{-1}\mathbf{O}^{\mathrm{T}} = \mathbf{A}^{-1}\mathbf{O}^{\mathrm{T}}, \tag{5.28}$$

and that the optimum efficiencies are given by

$$\xi_{\mathrm{opt},i}^2 = \frac{1}{n B_{ii}}, \tag{5.29}$$

where B_{ii} are the diagonal elements of $\mathbf{B} \equiv \mathbf{A}^{-1}$.

It is to be observed that the binding condition in Eq. (5.27) is more restrictive than that of Eq. (5.20), the former being a particular case of the latter. Therefore, $\xi_{\mathrm{opt},i}^2 \leq \xi_{\mathrm{max},i}^2$ or,

$$\frac{1}{B_{ii}} \leq \sum_{j=1}^{n} O_{ji}^2 = A_{ii}. \tag{5.30}$$

Equation (5.30) can then be read as *the greater the root mean square of the column elements of the modulation matrix, the greater the polarimetric efficiency attainable by the modulation scheme.*

Further reasoning leads to the conclusion that the equality sign holds for Eq. (5.30) only when matrix **A** is diagonal; hence, *the more diagonal matrix* **A** *is, the closer the polarimetric efficiency will be to maximum.* In other words, if two given modulation schemes have approximately the same A_{ii} elements, the one with

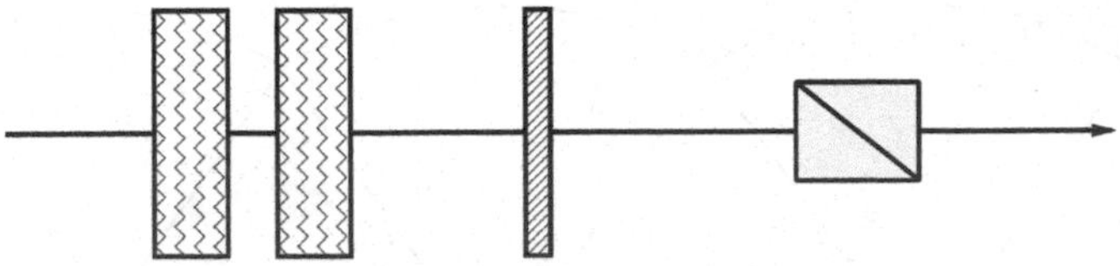

Fig. 5.4. Block diagram of ZIMPOL.

the relatively smaller non-diagonal elements is the more efficient. An application of these results to practical design is presented in Section 5.4.

5.3 Some solar Stokes polarimeters

Let us comment a little on *real* polarization analysis systems that are used in practice, namely, ZIMPOL (for Zürich IMaging POLarimeter; Stenflo *et al.*, 1992), ASP (for Advanced Stokes Polarimeter; Elmore *et al.*, 1992), TIP and LPSP (for Tenerife Infrared Polarimeter and La Palma Stokes Polarimeter; Martínez Pillet *et al.*, 1999). We outline here their main polarizing properties such as their block elements and their modulation matrices and efficiencies. We shall not enter into considerations of their performance. As a matter of fact, most of their differences stem from the various purposes they were designed for. Moreover, some may have evolved and others may have updated some block elements. Some of them may pursue ultimate polarimetric accuracy by perhaps sacrificing spatial resolution, whereas others may seek a balance between both, etc. Such appraisals of the possible aims of polarimeters is left to the interested reader.

5.3.1 ZIMPOL

The Zürich Imaging Polarimeter consists of two piezo-elastic modulators followed by a super-achromatic quarter-wave plate and a Glan polarizer (see Fig. 5.4). The piezo-elastic crystals modulate the diattenuation vector with oscillating retardances $\delta(t) + \pi/4$ and $\delta(t) - \pi/4$ of frequency 50 kHz and are oriented at 45° and 0° relative to the positive Stokes Q direction (the positive X direction in the discussion in Section 3.3). The fast axis of the quarter-wave plate is at 0° too and the optical axis of the polarizer is at 22.5°. After noticing the fast modulation frequencies, the reader may already have guessed that this polarimeter was devised to carry out pure time modulation.

A frequency of 50 kHz means ideal sampling intervals of 20 μs. These very short exposure times are impossible for conventional CCD cameras with their long read-out time (rows are read by shifting the charges through all the electronic wells). Moreover, if all polarimetric measurements are to be performed in the same pixel, an alternative detection system has to be conceived. Stenflo and coworkers

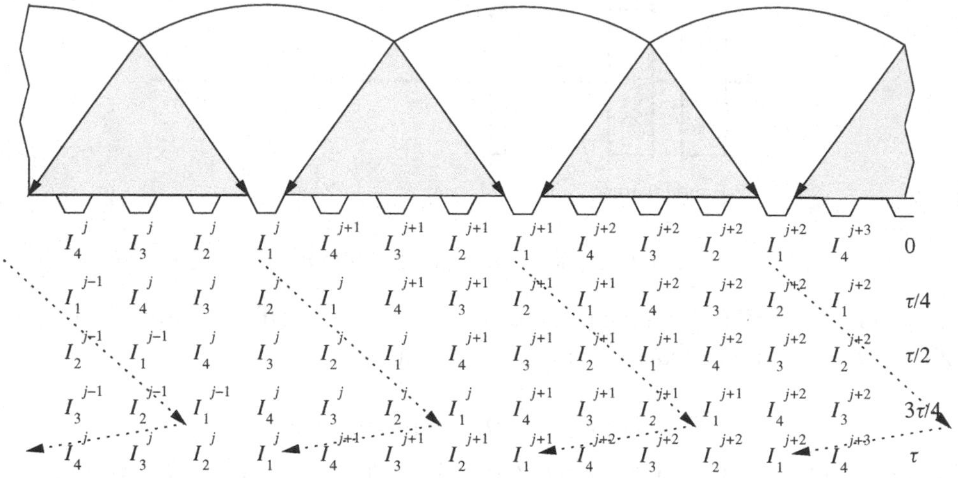

Fig. 5.5. Detection system for ZIMPOL. Three out of every four pixels in a row are masked. Charge is shifted forward by one pixel three times per cycle. In the last step, charge is shifted backwards by three pixels. I_i^j means $I^j_{\mathrm{meas},i}$, according to the convention mentioned in the text.

have succeeded in tailoring a CCD chip in a very clever way that is illustrated in Fig. 5.5. Imagine that the chip has n pixels per row. Three such pixels out of every four are masked so that they are not illuminated at any time. During the sampling interval, τ, every illuminated pixel j ($j = 1, 2, \ldots, n/4$) accumulates a given polarimetric signal, $I^j_{\mathrm{meas},i}$ ($i = 1, 2, 3, 4$), and a very short time (of order 0.3 μs) is allowed between sampling steps to shift the charges to the masked pixels in one direction (say forwards) but *without reading the chip*. In the last step of the cycle, charge is moved backwards. The three forward shifts are in steps of one pixel while the backward shift is three pixels in length, and all of them are carried out in synchrony with the piezo-elastic modulators. In this way, the cycle can be repeated until the desired S/N is achieved when the detector is immediately read out. Evidently, the backward shift implies that every pixel is exposed for 0.3 μs with an erroneous polarimetric state. This effect introduces very small errors that are nonetheless corrected by calibration. In order not to waste three-quarters of the available photons, the chip is covered with a micro-lens array that concentrates light onto the illuminated pixel, leaving the masked pixels shadowed.

A modulation matrix of ZIMPOL is

$$\mathbf{O}_{\mathrm{ZIMPOL}} = \begin{pmatrix} 1 & 0.39 & -0.66 & 0.26 \\ 1 & -0.01 & 0 & -0.40 \\ 1 & 0.39 & 0.66 & 0.26 \\ 1 & -0.77 & 0 & 0.92 \end{pmatrix}, \tag{5.31}$$

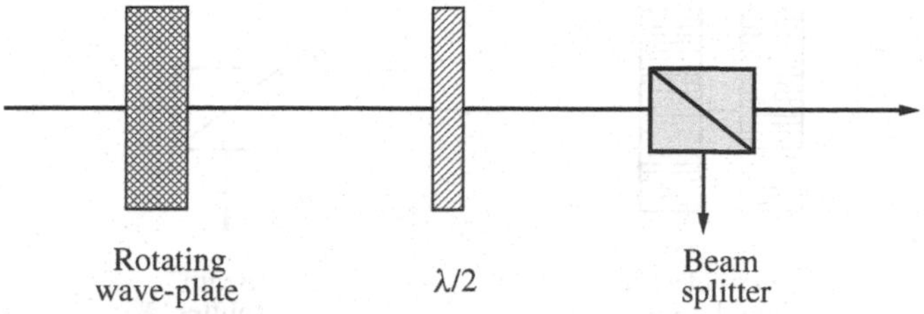

Fig. 5.6. Block diagram for ASP.

its polarimetric efficiency vector is

$$\boldsymbol{\xi} = (0.83, 0.39, 0.47, 0.38)^{\mathrm{T}}, \tag{5.32}$$

and the total efficiency for the polarization parameters is

$$\sqrt{\xi_2^2 + \xi_3^2 + \xi_4^2} = 0.72. \tag{5.33}$$

5.3.2 ASP

Besides some feed and beam steering optics, the Advanced Stokes Polarimeter consists of a rotating retarder of retardance $\pi/2$ at 740 nm, 107.28° at 630 nm, and 130.68° at 517 nm, an achromatic half-wave plate, and a beam splitter (see Fig. 5.6). The rotation of the retarder is performed at 3.75 Hz and the synchronized detection is made with specially tailored 12 bit CCD cameras working at video rates (60 Hz). The two output beams are measured so that a spatio–temporal modulation is carried out in the end.

The modulation cycle is such that eight polarimetric measurements are used in determining all four Stokes parameters. Hence, two polarimetric cycles are performed per second. The modulation matrix for each detector is

$$\mathbf{O}_{\mathrm{ASP}} = \begin{pmatrix} 1 & 0.77 & 0.41 & -0.36 \\ 1 & -0.06 & 0.41 & -0.86 \\ 1 & -0.06 & -0.41 & -0.86 \\ 1 & 0.77 & -0.41 & -0.36 \\ 1 & 0.77 & 0.41 & 0.36 \\ 1 & -0.06 & 0.41 & 0.86 \\ 1 & -0.06 & -0.41 & 0.86 \\ 1 & 0.77 & -0.41 & 0.36 \end{pmatrix}, \tag{5.34}$$

the polarimetric efficiency vector is

$$\boldsymbol{\xi}_{\mathrm{ASP}} = (0.76, 0.42, 0.41, 0.66), \tag{5.35}$$

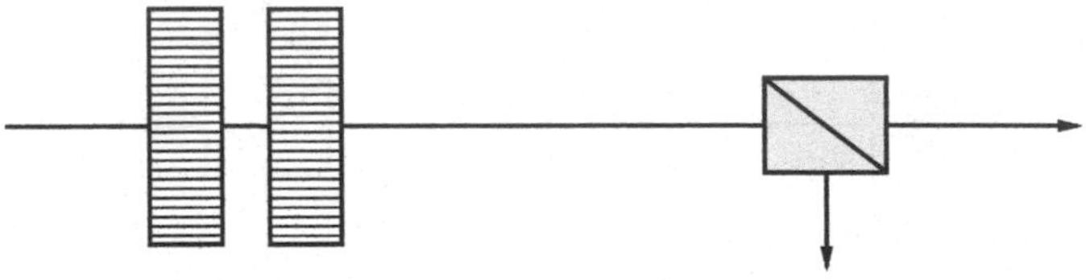

Fig. 5.7. Block diagram for TIP and LPSP.

and the total efficiency for the polarization parameters,

$$\sqrt{\xi_2^2 + \xi_3^2 + \xi_4^2} = 0.88. \tag{5.36}$$

5.3.3 *TIP and LPSP*

These two polarimeters are very similar. Their main difference is the wavelengths at which they are designed to observe: TIP is characterized for the near infrared (around 1.56 μm) and LPSP for visible wavelengths (around 630 nm). This difference led to small changes in the optical design and in the modulation cycle but both systems can be conceptually considered the same polarimeter. A diagram illustrating the analyzer can be seen in Fig. 5.7. It consists of two ferroelectric liquid crystals and a beam splitter. Spatio–temporal modulation is performed.

For both polarimeters, the (fixed) retardances of the two liquid crystals are close to a half wave and close to a quarter wave, respectively, at the nominal wavelengths. The difference in orientation angles of the optical axis between the two states of both crystals is 45° and their relative orientation is left as a free parameter that can be set by computer control, as can the exposure time of the CCD cameras, which are the same as those of the ASP. The orientation angles and differences between states are temperature sensitive so they are thermalized to the required temperatures. In the standard modulation scheme both use four polarimetric measurements of exposure time 16 ms for LPSP (15 Hz modulation frequency) and 50 ms (2 Hz modulation frequency) for TIP.

A modulation matrix for TIP is

$$\mathbf{O}_{\rm TIP} = \begin{pmatrix} 1 & 0.47 & -0.68 & 0.48 \\ 1 & -0.91 & -0.19 & -0.13 \\ 1 & -0.11 & 0.57 & 0.72 \\ 1 & 0.68 & 0.27 & -0.58 \end{pmatrix}, \tag{5.37}$$

the corresponding efficiency vector is

$$\xi_{\mathrm{TIP}} = (0.97, 0.61, 0.47, 0.51), \tag{5.38}$$

and the total efficiency for the polarization parameters is

$$\sqrt{\xi_2^2 + \xi_3^2 + \xi_4^2} = 0.92. \tag{5.39}$$

5.4 A practical discussion of polarimetric efficiencies

To illustrate the usefulness in practical design of the results obtained in Sections 5.2.1 and 5.2.2, let us digress a little on the actual efficiencies of the above polarimeters, their maximum attainable efficiencies and their possible improvements as far as polarimetric accuracy is concerned. It is important to note that the improvements we suggest in what follows may or may not be applied in reality, since some of them could imply modifications of some of the physical block components. These suggestions are made only to stress the importance of considering polarimetric accuracy from the very first steps in design. Deviations from optimum values may result from the compromises to which every polarimeter is subject.

According to inequality (5.30) the maximum attainable efficiencies of the three polarimeters are given by

$$\begin{aligned} \xi_{\mathrm{max,ZIMPOL}} &= (1.000, 0.474, 0.467, 0.534), \\ \xi_{\mathrm{max,ASP}} &= (1.000, 0.546, 0.410, 0.659), \\ \xi_{\mathrm{max,TIP}} &= (1.000, 0.617, 0.473, 0.525). \end{aligned} \tag{5.40}$$

A comparison of Eqs (5.40) with Eqs (5.32), (5.35), and (5.38) readily draws attention to four relevant features. First, all three polarimeters behave ideally for Stokes I and can reach unit efficiency for the total intensity. Second, TIP can (and in practice does) have higher efficiencies for Stokes Q and U than ASP and ZIMPOL. Third, ASP can (and in practice does) have higher polarimetric efficiency for Stokes V than ZIMPOL and TIP. And fourth, ZIMPOL and TIP can reach almost the same efficiency for Stokes U.

The actual value of the efficiencies for Stokes I are of the same order as the relative weight of non-diagonal to diagonal elements in the first column of their **A** matrices: 0.260, 0.355, 0.127. The same applies to the differences between $\xi_{2,\mathrm{ZIMPOL}}$ and $\xi_{2,\mathrm{TIP}}$; the relative weights of non-diagonal elements are 0.42, 0.07. This is in agreement with the second rule derived in Section 5.2.2: the more diagonal matrix **A** is, the closer the actual efficiency is to maximum.

According to the first rule of Section 5.2.2, the efficiencies could be raised by increasing the magnitude of the vector columns of matrix **O**. Let us find simple ways (feasible or otherwise in practice) of achieving this goal for ZIMPOL and ASP.

Imagine the ZIMPOL modulation cycle to comprise eight measurements: the four actual ones plus four others so that its modulation matrix were to become

$$\mathbf{O}_{\text{ZIMPOL,new}} = \begin{pmatrix} 1 & 0.39 & -0.66 & 0.26 \\ 1 & -0.01 & 0 & -0.40 \\ 1 & 0.39 & 0.66 & 0.26 \\ 1 & -0.77 & 0 & 0.92 \\ 1 & -0.39 & -0.66 & -0.26 \\ 1 & 0.01 & 0 & 0.40 \\ 1 & -0.39 & 0.66 & -0.26 \\ 1 & 0.77 & 0 & -0.92 \end{pmatrix}. \tag{5.41}$$

The new polarimetric efficiency vector would then be

$$\xi_{\text{ZIMPOL,new}} = (1.00, 0.41, 0.47, 0.46). \tag{5.42}$$

Imagine ASP to have the slightly different modulation matrix

$$\mathbf{O}_{\text{ASP,new}} = \begin{pmatrix} 1 & 0.77 & 0.41 & -0.36 \\ 1 & -0.06 & 0.41 & -0.86 \\ 1 & -0.06 & -0.41 & 0.86 \\ 1 & 0.77 & -0.41 & 0.36 \\ 1 & -0.77 & 0.41 & 0.36 \\ 1 & 0.06 & 0.41 & 0.86 \\ 1 & 0.06 & -0.41 & -0.86 \\ 1 & -0.77 & -0.41 & -0.36 \end{pmatrix}. \tag{5.43}$$

The new efficiencies would then be given by

$$\xi_{\text{ASP,new}} = (1.00, 0.55, 0.41, 0.66). \tag{5.44}$$

The new polarimetric efficiencies turn out to be higher than the actual ones because the new **A** matrices are more diagonal. As a matter of fact, in the case of ASP, the new **A** matrix is exactly diagonal and the efficiencies reach their maxima.

Recommended bibliography

Bianda, M., Solanki, S.K., and Stenflo, J.O. (1998). Hanle depolarisation in the solar chromosphere. *Astron. Astrophys.* **331**, 760.

Collados, M. (1999). High resolution spectropolarimetry and magnetometry, in *Third Advances in Solar Physics Euroconference: magnetic fields and oscillations*. B. Schmieder, A. Hofman, and J. Staude (eds.). ASP Conf. Ser. **184**, 3 (Publications of the Astronomical Society of the Pacific: San Francisco).

Donati, J.-F., Semel, M., Rees, D.E., Taylor, K., and Robinson, R.D. (1990). Detection of a magnetic region on HR 1099. *Astron. Astrophys.* **232**, L1.

Elmore, D.F., Lites, B.W., Tomczyk, S., Skumanich, A.P., Dunn, R.B., Schenke, J.A., Streander, K.V., Leach, T.W., Chambellan, C.W., Hull, H.K., and Lacey, L.B. (1992). The Advanced Stokes Polarimeter: a new instrument for solar magnetic field research. *Proc. Soc. Photo-Opt. Instrum. Eng.* **1746**, 22.

Gandorfer, A.M. and Povel, H.P. (1997). First observations with a new imaging polarimeter. *Astron. Astrophys.* **328**, 381.

Gray, D.F. (1992). *The observation and analysis of stellar photospheres*, 2nd Edition (Cambridge University Press; Cambridge). Chapter 3.

Keller, C.U., Bernasconi, P.N., Egger, U., Povel, H.P., and Stenflo, J.O. (1995). Visible and near-infrared polarimetry with LEST. *LEST Foundation Tech. Report* **59**. O. Engvold and Ø. Hauge (eds.) (The Institute of Theoretical Astrophysics: Oslo).

Lites, B.W. (1987). Rotating waveplates as polarization modulators for Stokes polarimetry on the sun: evaluation of seeing-induced crosstalk errors. *Appl. Opt.* **26**, 3838.

Lu, S.-Y. and Chipman, R.A. (1996). Interpretation of Mueller matrices based on polar decomposition. *J. Opt. Soc. Am. A* **13**, 1106.

Martínez Pillet, V. (1992). *Relations between fundamental parameters of sunspots*. PhD Thesis (Universidad de La Laguna: La Laguna).

Martínez Pillet, V., Collados, M., Sánchez Almeida, J., González, V., Cruz-López, A., Manescau, A., Joven, E., Páez, E., Díaz, J.J., Feeney, O., Sánchez, V., Scharmer, G., and Soltau, D. (1999). LPSP and TIP: full Stokes polarimeters for the Canary Islands observatories, in *High resolution solar physics: theory, observations, and techniques*. T. Rimmele, R. Raddick, and K. Balasubramaniam (eds.) ASP Conf. Ser. **184**, 264. (Publications of the Astronomical Society of the Pacific: San Francisco).

Mein, P. and Rayrole, J. (1985). THEMIS solar telescope. *Vistas in Astronomy* **28**, 567.

Mihalas, D. (1978). *Stellar atmospheres* (W.H. Freeman and Company: San Francisco). Chapter 1, Section 1.1.

Sánchez Almeida, J. and Martínez Pillet, V. (1992). Instrumental polarization in the focal plane of telescopes. *Astron. Astrophys.* **260**, 543.

Sánchez Almeida, J., Martínez Pillet, V., and Wittmann, A. (1991). The instrumental polarization of a Gregory-Coudé telescope. *Solar Phys.* **134**, 1.

Schröter, E.H., Soltau, D., and Wiehr, E. (1985). The German solar telescopes at the Observatorio del Teide. *Vistas in Astronomy* **28**, 519.

Semel, M., Donati, J.-F., and Rees, D.E. (1993). Zeeman-Doppler imaging of active stars. III. Instrumental and technical considerations. *Astron. Astrophys.* **278**, 231.

Stenflo, J.O. (1991). Optimization of the LEST polarization modulation system. *LEST Foundation Tech. Report* **44**. O. Engvold and Ø. Hauge (eds.) (The Institute of Theoretical Astrophysics: Oslo).

Stenflo, J.O. (1994). *Solar magnetic fields. Polarized radiation diagnostics* (Kluwer Academic Publishers: Dordrecht). Chapter 13.

Stenflo, J.O., Keller, C.U., and Povel, H.P. (1992). Demodulation of all four Stokes parameters with a single CCD. ZIMPOL II Conceptual design. *LEST Foundation Tech. Report* **54**, O. Engvold and Ø. Hauge (eds.) (The Institute of Theoretical Astrophysics: Oslo).

Stenflo, J.O. and Povel, H.P. (1985). Astronomical polarimeter with 2-D detector arrays. *Appl. Opt.* **24**, 3893.
del Toro Iniesta, J.C. and Collados, M. (2000). Optimum modulation and demodulation matrices for solar polarimetry. *Appl. Opt.* **39**, 1637.

6
Absorption and dispersion

"El clero era absorbente". Sobre todo Don Fermín había sido un poco jesuita.
—Leopoldo Alas, Clarín, 1885.

'The clergy were like a sponge.' And what was more, Don Fermin had once been something of a Jesuit.

So far we have avoided detailed discussion about two physical phenomena that are crucial in the context of this book and for any understanding of the interaction between matter and radiation in general. These two phenomena are absorption and dispersion, that is, the removal of energy from the electromagnetic field by matter and the dephasing of the electric field components as light streams through the medium. Although we have barely mentioned the existence of these effects, we shall need a deeper insight into both of them. We shall see that retardance, birefringence, and absorption properties of polarization systems, assumed in the preceding sections, are based on these phenomena, whose wavelength dependence is understood in terms of the wavelength dependence of the dielectric permittivity and, hence, of the refractive index of the medium. By studying absorption and dispersion we are producing the necessary bricks with which to build a theory of radiative energy transfer which will be discussed in following chapters. We shall continue to assume unit isotropic magnetic permeability of $\mu = 1$ for the medium.

Certainly, a full account of absorption and dispersion processes can be carried out only within the framework of quantum mechanics. Nonetheless, a semi-classical approach turns out to be adequate for grasping the underlying physics. This means that we shall stick to classical physics as much as possible, but that we shall resort occasionally to concepts or results from quantum mechanics.

6.1 Light propagation through low-density, weakly conducting media

From Maxwell's equations and their related constitutive (or material) equations, the only way for matter to draw energy from the electromagnetic field is through the appearance of Joule currents ($\boldsymbol{j} = \boldsymbol{\sigma E}$), which dissipate radiative energy in the form of heat. Note that a conductivity tensor, $\boldsymbol{\sigma}$, has been introduced because the medium is generally anisotropic.

Strictly speaking, the wave equation (2.1) and its analogue for anisotropic media [Eq. (4.5)] have a real coefficient $\varepsilon/c^2 = n^2/c^2$ only for non-conducting media ($\boldsymbol{\sigma} = \mathbf{0}$), as we are about to see. When the electrical conductivity is non-negligible, as is the case for a stellar plasma, the wave equation becomes†

$$\nabla^2 E_\alpha - \frac{\varepsilon_\alpha}{c^2}\ddot{E}_\alpha - \frac{4\pi\sigma_\alpha}{c^2}\dot{E}_\alpha = 0, \tag{6.1}$$

where $\alpha = 1, 2, 3$ applies to the three principal directions $(\hat{\boldsymbol{e}}_1, \hat{\boldsymbol{e}}_2, \hat{\boldsymbol{e}}_3)$ of the anisotropic medium.

Now, Fourier transforming Eq. (6.1) twice, first in space and then in time, gives‡

$$-k_\alpha^2 \bar{E}_\alpha + \frac{\varepsilon_\alpha \omega^2}{c^2}\bar{E}_\alpha + \mathrm{i}\frac{4\pi\sigma_\alpha\omega}{c^2}\bar{E}_\alpha = 0, \tag{6.2}$$

where $\bar{E}_\alpha$ denotes the double Fourier transform of E_α. Since Eq. (6.2) must be valid for every $\bar{E}_\alpha$, we conclude that the square wave number must be

$$k_\alpha^2 = \frac{\omega^2}{c^2}\left(\varepsilon_\alpha + \mathrm{i}\frac{4\pi\sigma_\alpha}{\omega}\right). \tag{6.3}$$

Equation (6.3) is known as the dispersion relation of the wave and is to be compared with its analogue for a non-conducting medium,

$$k_\alpha^2 = \frac{\omega^2}{c^2}\varepsilon_\alpha. \tag{6.4}$$

Thus, a complex wave number makes its appearance simply because of the presence of conductive properties in the medium.

The simplest solution of Eq. (6.1) is a purely monochromatic, time-harmonic plane wave of components

$$E_\alpha = a_\alpha\, \mathrm{e}^{\mathrm{i}\left[k_\alpha(\boldsymbol{r}\cdot\hat{\boldsymbol{t}}) - \omega t + \varpi_\alpha\right]} \tag{6.5}$$

that are identical to those of Eq. (2.9), except that now the vector wave number is complex, and that we specify here the ray direction, $\hat{\boldsymbol{t}}$, instead of the wavefront

† We assume that the conductivity tensor is diagonal when the dielectric tensor is.

‡ Note that the definition of Fourier transform used to obtain Eq. (6.2) is $\mathcal{F}[f] \equiv \int_{-\infty}^{\infty} f(t)\,\mathrm{e}^{\mathrm{i}\omega t}\,\mathrm{d}t$, consistent with our convention of considering $\mathrm{e}^{-\mathrm{i}\omega t}$ as the elementary harmonic function.

normal, $\hat{\boldsymbol{s}}$. This result implies that everything, including the generalization to polychromatic light, stays formally the same as in the non-conducting case but with a *complex* refractive index,

$$n_\alpha^2 \equiv \varepsilon_\alpha + \mathrm{i}\frac{4\pi\sigma_\alpha}{\omega}. \tag{6.6}$$

If we write

$$n_\alpha = 1 + \delta_\alpha + \mathrm{i}\kappa_\alpha \tag{6.7}$$

and assume that both δ_α and κ_α are real and much less than unity, then the dielectric and conductivity tensor components can safely be regarded as if they were real.† This assumption seems reasonable for a stellar atmosphere, where, on the one hand, densities are low enough for the real part of the refractive index to barely differ from the *in vacuo* unit value, and, on the other hand, although conductivity is non-negligible, its value is sufficiently small. With this assumption, neglecting second-order terms,

$$n_\alpha^2 \simeq 1 + 2(\delta_\alpha + \mathrm{i}\kappa_\alpha). \tag{6.8}$$

The components of a quasi-monochromatic plane wave traveling through the anisotropic medium become

$$E_\alpha(t) = \left\{ \mathcal{E}_\alpha(t)\, \mathrm{e}^{-\frac{\omega}{c}\kappa_\alpha (\boldsymbol{r}\cdot\hat{\boldsymbol{t}})} \mathrm{e}^{\mathrm{i}\left[\frac{\omega}{c}(1+\delta_\alpha)(\boldsymbol{r}\cdot\hat{\boldsymbol{t}}) + \varpi_\alpha(t)\right]} \right\} \mathrm{e}^{-\mathrm{i}\omega t}. \tag{6.9}$$

The imaginary part of the refractive index clearly produces a *real* decay factor that makes the amplitude of the wave decrease exponentially, thus accounting for electromagnetic energy removal. The real part of n_α takes care of shifting the phase across the ray path from its original value, ϖ_α.

If we do not consider these extinction and phase factors to be implicitly included in Eq. (3.6), then the coherency matrix of the light beam should be modified. In particular, if the ray direction coincides with one of the principal directions of the medium, say $\hat{\boldsymbol{e}}_3$, the diagonal elements of **C** are those of the non-conducting medium multiplied by

$$\mathrm{e}^{-\frac{1}{2}\frac{\omega}{c}(2\kappa_1+2\kappa_2)(\boldsymbol{r}\cdot\hat{\boldsymbol{t}})}, \tag{6.10}$$

and the non-diagonal elements of **C** are those of the non-conducting medium multiplied by

$$\mathrm{e}^{-\frac{1}{2}\frac{\omega}{c}(2\kappa_1+2\kappa_2)(\boldsymbol{r}\cdot\hat{\boldsymbol{t}})}\mathrm{e}^{\frac{\mathrm{i}}{2}\frac{\omega}{c}(2\delta_1-2\delta_2)(\boldsymbol{r}\cdot\hat{\boldsymbol{t}})}, \tag{6.11}$$

† ε_α and σ_α can indeed be complex for some media and frequencies.

and its complex conjugate, respectively. That is, matrix **C** of Eq. (3.11) becomes

$$\mathbf{C} = \mathrm{e}^{-\frac{1}{2}\frac{\omega}{c}(2\kappa_1+2\kappa_2)\,(\boldsymbol{r}\cdot\hat{\boldsymbol{t}})} \begin{pmatrix} \langle \mathcal{E}_1(t)\rangle^2 & \langle \mathcal{E}_1(t)\mathcal{E}_2(t)\mathrm{e}^{\frac{\mathrm{i}}{2}\upsilon}\rangle \\ \langle \mathcal{E}_1(t)\mathcal{E}_2(t)\mathrm{e}^{-\frac{\mathrm{i}}{2}\upsilon}\rangle & \langle \mathcal{E}_2(t)\rangle^2 \end{pmatrix}, \tag{6.12}$$

where

$$\upsilon \equiv \left[\frac{\omega}{c}(2\delta_1 - 2\delta_2)(\boldsymbol{r}\cdot\hat{\boldsymbol{t}}) + (2\varpi_1 - 2\varpi_2)\right]. \tag{6.13}$$

The coefficient

$$\chi_\alpha \equiv \frac{2\omega}{c}\kappa_\alpha = \frac{4\pi}{\lambda}\kappa_\alpha \tag{6.14}$$

is known as the *absorption coefficient* for light polarized in the α direction. The coefficient

$$\tilde{\chi}_\alpha \equiv \frac{2\omega}{c}\delta_\alpha = \frac{4\pi}{\lambda}\delta_\alpha \tag{6.15}$$

is known as the *dispersion coefficient* for light polarized in the α direction. It should be observed that polarization in the α direction means elliptical polarization in general, for the principal directions can be marked by complex vectors such as those in Section 3.6.

In summary, we may say that a stellar plasma is able to absorb radiative energy, because it is a conducting medium, in very much the same way as metals do. Furthermore, the negligible average charge density prevailing in the plasma is analogous to the absence of free charges in metals. This last property can be shown to be a direct consequence of Maxwell's equations.

From the exponential decay factor, we see that the intensity of the α component of the electric field decreases to a value $1/\mathrm{e}$ smaller than the original intensity over a length known as the *mean free path* of light polarized in the α direction,

$$\ell_\alpha \equiv \frac{1}{\chi_\alpha} = \frac{\lambda}{4\pi\kappa_\alpha}. \tag{6.16}$$

For solar atmospheres and visible wavelengths, the evaluation of the absorption coefficient gives a mean free path for photons of the order of 100 km, which in turn gives an order of magnitude of 4×10^{-13} for the imaginary part of the refractive index. Such a value fully justifies the above assumption of small κ_α and, on the other hand, is to be compared to that of metals in the laboratory, for which κ_α can be of the order of 10^3: in metals, the penetration depth of photons is very short; in fact, a perfect conductor ($\sigma_\alpha \longrightarrow \infty$) has $\ell_\alpha = 0$, and hence all the incident electromagnetic energy is reflected and cannot be transmitted through the medium.

Two final remarks are in order at this point. First, we have been assuming in this section that the medium is anisotropic and that in the frame of principal directions both the permittivity tensor, $\boldsymbol{\varepsilon}$, and the conductivity tensor, $\boldsymbol{\sigma}$, are diagonal.

Consequently, the refractive index, and the absorption and dispersion coefficients depend on the direction of propagation. For many astrophysical applications these conditions are too general and one can safely assume that the plasma has isotropic dielectric and conductive properties so that a single (complex) refractive index and single absorption and dispersion coefficients provide a coherent picture of the interaction between matter and radiation. Second, we have been dealing with dielectric and conductivity components without taking into consideration their wavelength dependence. This is a final ingredient, which is discussed in the following section.

6.2 Absorption and dispersion profiles

The relationship (4.2) between $\boldsymbol{D}$ and $\boldsymbol{E}$ implicitly regards the interaction between radiation and matter as though the electric field produces at every point in the medium displacements that have components proportional to the components of $\boldsymbol{E}$ in the frame of principal directions. After the discussion of the preceding section, let us replace $\boldsymbol{\varepsilon}$ by $\boldsymbol{\varepsilon} + \mathrm{i}(4\pi/\omega)\boldsymbol{\sigma}$. If the medium is assumed to be made of atoms, i.e., negative electrons surrounding positive nuclei, the individual displacements can be thought of as the creation of electric dipoles,

$$\boldsymbol{p} = -\mathrm{e}_0\boldsymbol{r}, \tag{6.17}$$

where $-\mathrm{e}_0$ is the charge of the electron and $\boldsymbol{r}$ the vector position of the electron motion induced by the external field. If N is the number volume density of such dipoles, the *electric polarization* vector,

$$\boldsymbol{P} = -N\mathrm{e}_0\boldsymbol{r}, \tag{6.18}$$

represents the total effect per unit volume.

The overall electric displacement, $\boldsymbol{D}$, can be written as

$$\boldsymbol{D} = \boldsymbol{E} + 4\pi\boldsymbol{P}, \tag{6.19}$$

where the factor 4π accounts for all possible directions of impinging radiation. Hence, $\boldsymbol{D}$ depends on the vector position of the electrons through

$$\boldsymbol{D} = \boldsymbol{E} - 4\pi N\mathrm{e}_0\boldsymbol{r}. \tag{6.20}$$

Calculation of the individual displacements can be carried out classically through the Lorentz electron theory.† In this phenomenological theory, the electron is seen as a superposition of classical oscillators,

$$\boldsymbol{r}(t) = \boldsymbol{d}(t)\ \mathrm{e}^{\mathrm{i}\omega t}, \tag{6.21}$$

† Further generalizations resulting from the quantum theory are also needed and will be introduced later.

where now $\boldsymbol{d}$ is the time-dependent, complex amplitude of the motion. The oscillators are excited by the force associated with the external electric field,

$$\boldsymbol{F}_{\mathrm{e}}(t) = -\mathrm{e}_0 \boldsymbol{E}(t),$$

where $\boldsymbol{E}(t)$ is given by Eq. (6.9). The oscillators are restored by a quasi-elastic force, which binds them in an equilibrium position,

$$\boldsymbol{F}_{\mathrm{r}}(t) = -q\boldsymbol{r}(t),$$

and they are damped by a resisting force,

$$\boldsymbol{F}_{\mathrm{d}}(t) = -m\gamma\dot{\boldsymbol{r}}(t),$$

where m is the rest mass of the electron. The damping force is indispensable in accounting for dissipative effects. Among these effects, one can enumerate absorption, which supplies heat (kinetic energy) to the particles, electromagnetic wave emission by the oscillating charges that carry energy away, scattering of photons in directions other than the original, and inelastic collisions between the absorbers and other particles in the medium.‡

Note that a "polychromatic" oscillator has been assumed because the exciting electric field is also polychromatic. The mathematical expression of Eq. (6.21) is analogous to, and in fact has the same justification as, Eq. (3.6) for the quasi-monochromatic beam of light. Most textbooks use a monochromatic approach instead.

Let us choose the system of complex unit vectors $\{\hat{\boldsymbol{e}}_{+1}, \hat{\boldsymbol{e}}_0, \hat{\boldsymbol{e}}_{-1}\}$ as the principal reference frame, so that

$$\begin{aligned} \hat{\boldsymbol{e}}_{\pm 1} &\equiv \frac{1}{\sqrt{2}}(\hat{\boldsymbol{e}}_1 \pm \mathrm{i}\hat{\boldsymbol{e}}_2) \\ \hat{\boldsymbol{e}}_0 &\equiv \hat{\boldsymbol{e}}_3. \end{aligned} \tag{6.22}$$

From a classical point of view, the choice of such a reference frame can be interpreted as the decomposition of the electron oscillation into a linear combination of three oscillators: one linear oscillator along $\hat{\boldsymbol{e}}_0$ and two circular oscillators: one revolving clockwise around $\hat{\boldsymbol{e}}_0$ (as seen through $-\hat{\boldsymbol{e}}_0$; i.e., right-handed), corresponding to $\hat{\boldsymbol{e}}_{-1}$, and another revolving counter-clockwise (i.e., left-handed), corresponding to $\hat{\boldsymbol{e}}_{+1}$. From a quantum-mechanical point of view, this system is linked to the absorption and emission of photons whose state is a linear combination of the only three pure states characterized by the third component of their angular momenta in the atomic reference system, $m_j = +1, 0, -1$. These three pure quantum states are

‡ The interested reader is referred to classical textbooks on stellar atmospheres, e.g., Mihalas (1978) or Gray (1992), to retrieve the means for actual calculation of absorption and damping coefficients.

indeed linked to the three totally polarized states (left-handed circular, linear, and right-handed circular).

Newton's second law applied to the motion of the electron gives, for each of the three components r_α, $\alpha = +1, 0, -1$,

$$m\ddot{r}_\alpha(t) = -\mathrm{e}_0 E_\alpha(t) - q_\alpha r_\alpha(t) - m\gamma_\alpha \dot{r}_\alpha(t), \tag{6.23}$$

where we have again made the assumption that both the $\boldsymbol{q}$ and $\boldsymbol{\gamma}$ tensors are diagonal in the principal reference frame.

Fourier transforming Eq. (6.23) in time we obtain†

$$\bar{d}_\alpha = -\frac{\mathrm{e}_0}{m} \frac{\bar{\mathcal{A}}_\alpha}{(q_\alpha/m) - \omega^2 - \mathrm{i}\omega\gamma_\alpha}, \tag{6.24}$$

where $\bar{d}_\alpha$ and $\bar{\mathcal{A}}_\alpha$ represent the Fourier transforms of $d_\alpha(t)$ and $\mathcal{A}_\alpha(t)$, the latter being the time-dependent complex amplitude of the electric field of Eq. (6.9); i.e.,

$$E_\alpha(t) \equiv \mathcal{A}_\alpha(t)\, \mathrm{e}^{-\mathrm{i}\omega t}.$$

Since the proportionality factor between $\bar{d}_\alpha$ and $\bar{\mathcal{A}}_\alpha$ is independent of time, it is readily seen that

$$d_\alpha(t) = -\frac{\mathrm{e}_0}{m} \frac{\mathcal{A}_\alpha(t)}{(q_\alpha/m) - \omega^2 - \mathrm{i}\omega\gamma_\alpha}. \tag{6.25}$$

The term q_α/m in the above equation is identified as a square resonant frequency, $\omega^2_{0,\alpha}$, at which, in the absence of damping, the individual displacements behave as

$$\lim_{\substack{\omega\to\omega_{0,\alpha}\\ \omega<\omega_{0,\alpha}}} d_\alpha = -\infty$$

and

$$\lim_{\substack{\omega\to\omega_{0,\alpha}\\ \omega>\omega_{0,\alpha}}} d_\alpha = +\infty.$$

Certainly, the presence of damping eliminates the singularity.

Since $(\omega_{0,\alpha} - \omega)/\omega \ll 1$ for quasi-monochromatic light (we are dealing with frequencies close to the resonant frequency), the difference $(\omega^2_{0,\alpha} - \omega^2)$ can be approximated to first order by $2\omega(\omega_{0,\alpha} - \omega)$. Writing further $\Gamma_\alpha \equiv \gamma_\alpha/4\pi$, the individual displacement components can be recast in the form

$$r_\alpha(t) \simeq \frac{-\mathrm{e}_0}{8\pi^2 m\nu} \frac{E_\alpha(t)}{\nu_{0,\alpha} - \nu - \mathrm{i}\Gamma_\alpha} \tag{6.26}$$

and are seen to be proportional to the principal components of the exciting electric field. This result is consistent with the proportionality between the principal

† See footnote ‡ on page 88.

electric displacement and electric field components. In fact, Eq. (6.20) can now be written as

$$D_\alpha = \left(1 + \frac{N e_0^2}{2\pi m \nu} \frac{\nu_{0,\alpha} - \nu + \mathrm{i}\Gamma_\alpha}{(\nu_{0,\alpha} - \nu)^2 + \Gamma_\alpha^2}\right) E_\alpha. \tag{6.27}$$

Now, the coefficient of proportionality between D_α and E_α is for sure the square refractive index,

$$n_\alpha^2 = 1 + \frac{N e_0^2}{2\pi m \nu} \frac{\nu_{0,\alpha} - \nu + \mathrm{i}\Gamma_\alpha}{(\nu_{0,\alpha} - \nu)^2 + \Gamma_\alpha^2}. \tag{6.28}$$

After eliminating n_α^2 from Eqs (6.8) and (6.28) and identifying the real and imaginary parts, we finally get the following *real* expressions for δ_α and κ_α:

$$\begin{aligned} \delta_\alpha &= \frac{N e_0^2}{4\pi m \nu} \frac{\nu_{0,\alpha} - \nu}{(\nu_{0,\alpha} - \nu)^2 + \Gamma_\alpha^2}, \\ \kappa_\alpha &= \frac{N e_0^2}{4\pi m \nu} \frac{\Gamma_\alpha}{(\nu_{0,\alpha} - \nu)^2 + \Gamma_\alpha^2}. \end{aligned} \tag{6.29}$$

As discussed in Section 6.1, κ_α is responsible for the absorption and δ_α for the dispersion. The absorption coefficient then has the expression

$$\chi_\alpha(\nu) = \frac{N\pi e_0^2}{mc} \frac{\Gamma_\alpha/\pi}{(\nu_{0,\alpha} - \nu)^2 + \Gamma_\alpha^2} \equiv \frac{N\pi e_0^2}{mc} \phi_\alpha(\nu), \tag{6.30}$$

where the second factor on the right-hand side is a Lorentzian profile, $\phi_\alpha(\nu)$, normalized to unit area (see Fig. 6.1); i.e.,

$$\int_{-\infty}^{+\infty} \chi_\alpha(\nu) \mathrm{d}\nu = \frac{N\pi e_0^2}{mc} \int_{-\infty}^{+\infty} \phi_\alpha(\nu) \mathrm{d}\nu = \frac{N\pi e_0^2}{mc}. \tag{6.31}$$

The dispersion coefficient is expressed as

$$\tilde{\chi}_\alpha(\nu) = \frac{N\pi e_0^2}{mc} \frac{1}{\pi} \frac{\nu_{0,\alpha} - \nu}{(\nu_{0,\alpha} - \nu)^2 + \Gamma_\alpha^2} \equiv \frac{N\pi e_0^2}{mc} \psi_\alpha(\nu), \tag{6.32}$$

where the second factor on the right-hand side is a dispersion profile, $\psi_\alpha(\nu)$, whose antisymmetric shape is also illustrated in Fig. 6.1. Obviously, the integral of this profile over the entire spectrum is zero. For historical reasons, some texts speak of "anomalous" dispersion profiles, so that a "normal" dispersion profile would be that with a singularity at the resonant frequency. We have already commented that this "normal" dispersion can only take place in the absence of the otherwise unavoidable damping.

One can clearly see in Eqs (6.30) and (6.32) that, if the resonant frequencies and damping components are all three equal, the medium can be considered isotropic.

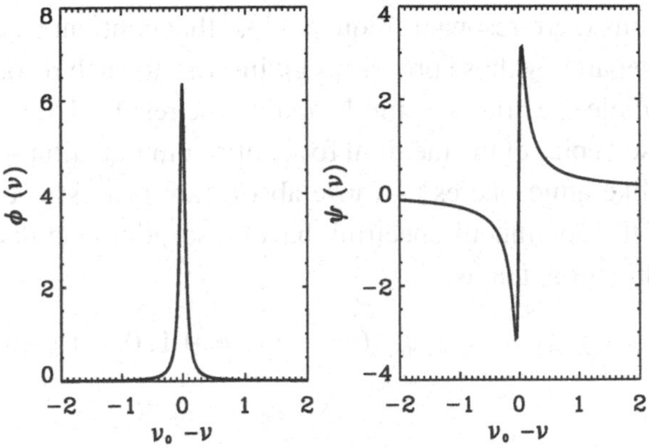

Fig. 6.1. Normalized absorption and dispersion profiles. Frequencies are in units of the Doppler width of the line (see Section 6.4). The damping coefficient is 0.05 in such units.

This is usually assumed for most cases of interest, and then the stellar atmosphere is characterized by a single index of refraction. The more general case has been discussed in here to stress first that isotropy is indeed an assumption (some mechanism might produce effects classically interpretable as refractive index anisotropies), and second that the presence of a magnetic field in the atmosphere does induce an anisotropy (a preferential direction), even when the damping components can be assumed to be isotropic ($\Gamma_\alpha = \Gamma, \forall\, \alpha = 1, 2, 3$).

Although fairly simple, it may be helpful to observe the slight formal changes that appear in the normalized absorption and dispersion profiles when they are considered as wavelength distributions instead of frequency distributions. Considering λ as the integration variable and assuming that $\lambda - \lambda_{0,\alpha} \ll \lambda_{0,\alpha}$ (as it is in practice), then

$$\phi_\alpha(\lambda) = \frac{1}{\pi} \frac{\Gamma_\alpha \lambda_{0,\alpha}^2 / c}{(\lambda - \lambda_{0,\alpha})^2 + \left(\Gamma_\alpha \lambda_{0,\alpha}^2 / c\right)^2} \tag{6.33}$$

and

$$\psi_\alpha(\lambda) = \frac{1}{\pi} \frac{\lambda - \lambda_{0,\alpha}}{(\lambda - \lambda_{0,\alpha})^2 + \left(\Gamma_\alpha \lambda_{0,\alpha}^2 / c\right)^2}. \tag{6.34}$$

6.3 A correction from quantum mechanics

Although just a single resonant frequency has been dealt with, it is evident that the medium can have as many resonances as those of its constituent atoms and molecules. These resonances may correspond to the frequencies needed to produce atomic transitions between bound electronic levels so that spectral lines are formed, or bound–free transitions, as in ionization or recombination processes, and free–free transitions (zero resonant frequency) so that continuous absorption takes place. Studying separately those processes giving rise to such resonances, we shall then model the whole spectrum by simply adding the results. Under the assumption of negligible anisotropies of the medium for continuum radiation – a likely approximation for sun-like atmospheres† – those absorption processes corresponding to the formation of the continuous spectrum have absorption and dispersion profiles independent of direction; that is,

$$\chi_{\rm cont}(\nu) = \chi_{{\rm cont},\alpha}(\nu), \quad \forall\, \alpha = +1, 0, -1 \tag{6.35}$$

and

$$\tilde{\chi}_{\rm cont}(\nu) = \tilde{\chi}_{{\rm cont},\alpha}(\nu), \quad \forall\, \alpha = +1, 0, -1. \tag{6.36}$$

Note that if Eqs (6.35) and (6.36) apply, continuum radiation cannot change its state of polarization after absorption and dispersion. Since the second factor in expression (6.11) is unity for continuum radiation, the output coherency matrix is proportional to the input coherency matrix:

$$\mathbf{C}'_{\rm cont} = {\rm e}^{-\chi_{\rm cont}\, \boldsymbol{r}\cdot\hat{\boldsymbol{t}}}\, \mathbf{C}_{\rm cont}. \tag{6.37}$$

Thus, all four Stokes parameters are multiplied by the same factor:

$$\boldsymbol{I}'_{\rm cont} = {\rm e}^{-\chi_{\rm cont}\, \boldsymbol{r}\cdot\hat{\boldsymbol{t}}}\, \boldsymbol{I}_{\rm cont}. \tag{6.38}$$

Therefore, if continuum radiation is unpolarized on input it then remains unpolarized on output. It is important to note that, within the limited range of frequencies of a given spectral line, the continuous absorption and dispersion profiles stay essentially constant, as can easily be deduced from their definitions. Hence, we can drop the explicit dependence on frequency and write $\chi_{\rm cont}$ and $\tilde{\chi}_{\rm cont}$.

Let us concentrate on spectral line formation, that is, on those processes producing transitions between two bound electronic states or between a bound state and the continuum. As a result of the wavelength dependence of the absorption processes, a spectral line has been formed for each polarization state in the principal directions of the medium whose shapes are given by the absorption profiles $\chi_\alpha(\nu)$, $\alpha = +1, 0, -1$. The absorption profiles account for the drawing of electromagnetic energy by the medium. In their turn, dispersion profiles explain the change in phase undergone by light after streaming through the medium. These results have been derived from the simple phenomenological theory of Lorentz. As a matter of fact, Lorentz's results are exact for electric dipole transitions when compared with more rigorous quantum-mechanical calculations, except for the frequency-integrated strength of the profiles [Eq. (6.31)]. The introduction is necessary of a

† We are in fact neglecting scattering polarization, which is observed close to the solar limb.

factor f, called the *oscillator strength*, which is proportional to the square modulus of the dipole matrix element between the lower and the upper levels involved in the transition. After this correction, the absorption and dispersion profiles become

$$\chi_\alpha(\nu) = \frac{N\pi e_0^2 f}{mc}\,\phi_\alpha(\nu) \tag{6.39}$$

and

$$\tilde{\chi}_\alpha(\nu) = \frac{N\pi e_0^2 f}{mc}\,\psi_\alpha(\nu). \tag{6.40}$$

6.4 Accounting for thermal motions in the medium

So far, we have implicitly assumed that material atoms and ions are at rest. Because of thermal agitation, however, even if the bulk or mean velocity in the direction of propagation (along the line of sight) is zero, every atom will have a non-zero velocity component. Let us assume that the distribution of velocities is Maxwellian, so that we shall have a number of ions with velocities in the vicinity of v given by

$$N(v) = \frac{N}{\sqrt{\pi}\,v_D}\,e^{-(v^2/v_D^2)}, \tag{6.41}$$

whose root-mean-square width or, *Doppler width*, v_D, is related to the temperature, T, of the medium by

$$v_D \equiv \sqrt{\frac{2kT}{m} + \xi_{mic}^2}, \tag{6.42}$$

where k is the Boltzmann constant and m the rest mass of the atom. We have introduced in Eq. (6.42) the *ad hoc* parameter ξ_{mic}, or *microturbulence velocity*, that is often used in astrophysics to account for those motions on smaller scales than the mean free path of photons which are not included in the Doppler width of the line. This parameter can simply be ignored for non-astrophysical applications. Being dependent on the mass of the absorber, the Doppler width is indeed a characteristic of each ionic species. Spectral lines of different atoms obviously have different Doppler velocities.

As a consequence of the motion, the atom "sees" the photons shifted in frequency because of the Doppler effect. The distribution of velocities can then be translated into a frequency distribution:

$$N(\nu) = \frac{N}{\sqrt{\pi}\,\Delta\nu_D}\,e^{-\left(\Delta\nu^2/\Delta\nu_D^2\right)} \equiv Np(\nu), \tag{6.43}$$

where

$$v = \frac{\Delta\nu\, c}{\nu_{0,\alpha}} = \frac{(\nu_{0,\alpha} - \nu)c}{\nu_{0,\alpha}}, \tag{6.44}$$

in conformance with the astrophysical convention that redshifts correspond to positive velocities, and

$$v_\mathrm{D} = \frac{\Delta\nu_\mathrm{D}\, c}{\nu_{0,\alpha}}. \tag{6.45}$$

Note that the distribution $p(\nu)$ is defined to have unit area so that the total population of absorbers is still N.

We must now take into account that the normalized absorption and dispersion profiles $\phi_\alpha(\nu)$ and $\psi_\alpha(\nu)$ of Eqs (6.39) and (6.40) were obtained for a set of absorbers with a distribution of frequency shifts equal to the Dirac $\delta(\nu)$ distribution, i.e., for a zero distribution of velocities. If absorption is a linear invariant against translations of the variable, continuous process, system theory clearly tells us that the normalized absorption and dispersion profiles for a distribution of frequency shifts, $p(\nu)$, are the convolutions of $\phi_\alpha(\nu)$ and $\psi_\alpha(\nu)$ with $p(\nu)$:

$$p(\nu) * \phi_\alpha(\nu) \quad \text{and} \quad p(\nu) * \psi_\alpha(\nu).$$

Alternatively, the distribution of velocities can be described by means of wavelength, rather than frequency, shifts. It is easily found that

$$N(\lambda) = \frac{N}{\sqrt{\pi}\,\Delta\lambda_\mathrm{D}}\,\mathrm{e}^{-(\Delta\lambda^2/\Delta\lambda_\mathrm{D}^2)} \equiv Np(\lambda), \tag{6.46}$$

where

$$v = \frac{\Delta\lambda\, c}{\lambda_{0,\alpha}} = \frac{(\lambda - \lambda_{0,\alpha})c}{\lambda_{0,\alpha}} \tag{6.47}$$

and

$$v_\mathrm{D} = \frac{\Delta\lambda_\mathrm{D}\, c}{\lambda_{0,\alpha}}. \tag{6.48}$$

The normalized absorption and dispersion profiles then turn out to be the convolutions

$$p(\lambda) * \phi_\alpha(\lambda) \quad \text{and} \quad p(\lambda) * \psi_\alpha(\lambda).$$

Before proceeding with convolutions, it is convenient to write our distributions in terms of the *reduced* variables

$$u_{0,\alpha} \equiv \frac{\nu_{0,\alpha} - \nu}{\Delta\nu_\mathrm{D}}, \tag{6.49}$$

or

$$u_{0,\alpha} \equiv \frac{\lambda - \lambda_{0,\alpha}}{\Delta\lambda_\mathrm{D}}, \tag{6.50}$$

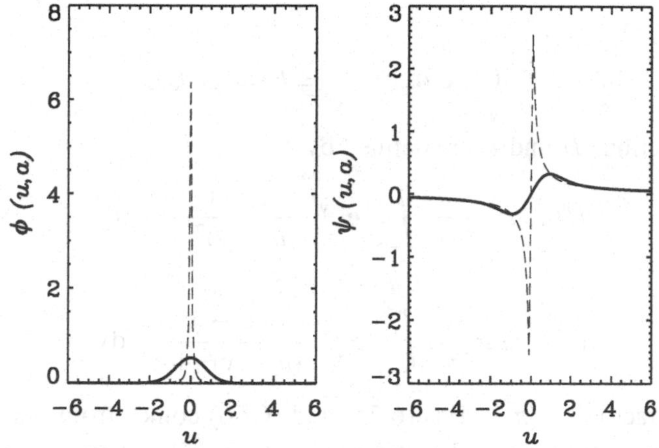

Fig. 6.2. An example of Voigt and Faraday–Voigt functions, which are the convolution with a Gaussian of the normalized absorption and dispersion profiles of Fig. 6.1.

depending on whether we are dealing with frequencies or wavelengths. With these definitions, one has†

$$p(u_{0,\alpha}) = \frac{1}{\sqrt{\pi}} \mathrm{e}^{-u_{0,\alpha}^2}, \tag{6.51}$$

$$\phi_\alpha(u_{0,\alpha}, a_\alpha) = \frac{1}{\pi} \frac{a_\alpha}{u_{0,\alpha}^2 + a_\alpha^2}, \tag{6.52}$$

and

$$\psi_\alpha(u_{0,\alpha}, a_\alpha) = \frac{1}{\pi} \frac{u_{0,\alpha}}{u_{0,\alpha}^2 + a_\alpha^2}, \tag{6.53}$$

where

$$a_\alpha \equiv \frac{\Gamma_\alpha}{\Delta\nu_\mathrm{D}} \quad \text{or} \quad a_\alpha \equiv \frac{\Gamma_\alpha \lambda_{0,\alpha}^2}{c\Delta\lambda_\mathrm{D}}. \tag{6.54}$$

Therefore, we can obtain single formal expressions valid for both the frequency and the wavelength treatments. Attention need only be paid to the choice for $u_{0,\alpha}$ and a_α. Keeping the old symbols for ϕ_α and ψ_α after the convolutions are carried out, the new normalized absorption and dispersion profiles are given by the Voigt and Faraday–Voigt functions (see Fig. 6.2):

$$\phi_\alpha(u_{0,\alpha}, a_\alpha) = \frac{1}{\sqrt{\pi}} H(u_{0,\alpha}, a_\alpha) \tag{6.55}$$

† A factor $1/\Delta\nu_\mathrm{D}$ appears explicitly in some texts (e.g., Landi Degl'Innocenti, 1992). One should understand that the integration variable in these texts is still the frequency even though use has been made of the variable u.

and

$$\psi_\alpha(u_{0,\alpha}, a_\alpha) = \frac{1}{\sqrt{\pi}} F(u_{0,\alpha}, a_\alpha), \tag{6.56}$$

where the functions H and F are defined by

$$H(u, a) \equiv \frac{a}{\pi} \int_{-\infty}^{\infty} e^{-y^2} \frac{1}{(u-y)^2 + a^2} \, dy \tag{6.57}$$

and

$$F(u, a) \equiv \frac{1}{\pi} \int_{-\infty}^{\infty} e^{-y^2} \frac{u-y}{(u-y)^2 + a^2} \, dy. \tag{6.58}$$

Note that the factor $1/\sqrt{\pi}$ in Eqs (6.55) and (6.56) comes from our definition of the Maxwellian distribution. Other definitions are possible. For example, if we take a Doppler width $\sqrt{\pi}$ times smaller than that of Eq. (6.42), this factor vanishes in Eqs (6.55) and (6.56).

As a consequence of thermal motions the normalized absorption and dispersion profiles have been considerably broadened as compared with those prevailing in the absence of such motions.† This means that significant absorption and dispersion processes may take place at far from central frequencies.

6.5 Spectral line absorption in moving media

A further generalization remains which concerns the mean macroscopic velocity of the medium in the propagation direction of light $v_{\rm LOS}$ (the index LOS stands for "line of sight"). Such a bulk motion of the medium shifts the frequency and wavelength of photons by

$$\Delta\nu_{\rm LOS} \equiv \nu_{0,\alpha} \frac{v_{\rm LOS}}{c} \quad \text{and} \quad \Delta\lambda_{\rm LOS} \equiv \lambda_{0,\alpha} \frac{v_{\rm LOS}}{c}. \tag{6.59}$$

The distribution of frequency shifts now turns out to be $p(\nu - \Delta\nu_{\rm LOS})$ and that of wavelength shifts $p(\lambda - \Delta\lambda_{\rm LOS})$ if we follow the astrophysical convention of positive velocities for redshifts. Note that there is no apparent change in sign between the two shift distributions. Because of the invariance against translations of the variable, the normalized absorption and dispersion profiles become

$$\phi_\alpha(u_{0,\alpha}, a_\alpha) = \frac{1}{\sqrt{\pi}} H(u_{0,\alpha} - u_{\rm LOS}, a_\alpha) \tag{6.60}$$

and

$$\psi_\alpha(u_{0,\alpha}, a_\alpha) = \frac{1}{\sqrt{\pi}} F(u_{0,\alpha} - u_{\rm LOS}, a_\alpha), \tag{6.61}$$

where, obviously, $u_{\rm LOS}$ is either $\Delta\nu_{\rm LOS}/\Delta\nu_{\rm D}$ or $\Delta\lambda_{\rm LOS}/\Delta\lambda_{\rm D}$.

† Pressure broadening produces Lorentzian and dispersion profiles like natural broadening (e.g., Gray, 1992).

We can now repeat Eqs (6.39) and (6.40) for the final expressions of the absorption and dispersion profiles,

$$\chi_\alpha(u_{0,\alpha}, a_\alpha) = \frac{N\pi e_0^2 f}{mc}\, \phi_\alpha(u_{0,\alpha}, a_\alpha) \tag{6.62}$$

and

$$\tilde{\chi}_\alpha(u_{0,\alpha}, a_\alpha) = \frac{N\pi e_0^2 f}{mc}\, \psi_\alpha(u_{0,\alpha}, a_\alpha), \tag{6.63}$$

where the normalized profiles are now given by Eqs (6.60) and (6.61).

Recommended bibliography

Born, M. and Wolf, E. (1993). *Principles of optics*, 6th edition (Pergamon Press: Oxford). Chapter 2. Section 2.3, Chapter 13. Sections 13.1–13.3.

Cox, J.P. and Giuli, R.T. (1968). *Principles of stellar structure. Vol. I: Physical principles* (Gordon and Breach, Science Publishers: New York). Chapter 2, Section 2.10.

Gray, D.F. (1992). *The observation and analysis of stellar photospheres*, 2nd edition (Cambridge University Press: Cambridge). Chapter 11.

Landi Degl'Innocenti, E. (1992). Magnetic field measurements, in *Solar observations: techniques and interpretations*. F. Sánchez, M. Collados, and M. Vázquez (eds.) (Cambridge University Press: Cambridge). Section 6.

Mihalas, D. (1978). *Stellar atmospheres*, 2nd edition. (W.H. Freeman and Co.: San Francisco). Chapter 9.

Stenflo, J.O. (1994). *Solar magnetic fields. Polarized radiation diagnostics* (Kluwer Academic Publishers: Dordrecht). Chapter 3.

7

The radiative transfer equation

Y entonces, el maestro sacó la daga, y dijo: —"Yo no sé quién es Ángulo ni Obtuso, ni en mi vida oí decir tales nombres; pero, con ésta en la mano, le haré yo pedazos".

—F. de Quevedo y Villegas, 1603?

And then the master drew his dagger and said, 'I never in my life heard of Angle or Obtuse, but with this in my hand I'll cut him to ribbons.'

So far we have been dealing with the propagation of light through media whose refractive indices have been assumed to be constant with position (the assumption of homogeneity). We are now able to embark on the study of the propagation of light through media whose refractive indices – and hence absorptive and dispersive properties – may vary along the ray path; a differential treatment is then in order. More specifically, we shall deal with stratified media whose material properties are constant in planes perpendicular to a given direction. Moreover, our study will not only include *passive* systems but emission properties of the medium will also be considered (although in the most simplified way).

There are three main hypotheses we should add to proceed with the development that follows:

(1) We shall assume that the absorptive, dispersive, and emissive properties of the medium are independent of the light-beam Stokes vector. This is in fact a linear approximation that holds in many astrophysical applications, where, even though the medium may be dependent on the whole radiation field, the angular width of the beam (indeed within the realm of geometrical optics) is so small that its contribution to the physical conditions of the medium can be neglected (e.g., Landi Degl'Innocenti and Landi Degl'Innocenti 1981).

(2) The radiation field will be assumed to be time-independent or, being more specific, the rate of changes in $\boldsymbol{I}$ is much slower than the relaxation time (or time for the photons to go through a mean free path). Time derivatives $\partial \boldsymbol{I}/\partial t$ are set to zero.

(3) Although the medium is inhomogeneous, the refractive index is so close to unity [Eq. (6.7)] that the effects of the term $\nabla\left[\boldsymbol{E}\cdot\nabla(\ln n^2)\right]$ on the wave equation can be neglected.†

7.1 A little geometry

Consider a ray of quasi-monochromatic light propagating along the Z axis of a medium with principal axes given by the (right-handed) orthonormal set of vectors $\{\hat{\boldsymbol{e}}_1, \hat{\boldsymbol{e}}_2, \hat{\boldsymbol{e}}_0\}$. It is important to note that we do not deal with the wavefront normal but with the ray direction $\hat{\boldsymbol{t}}\,(=\hat{\boldsymbol{z}})$. The observation direction (the line of sight in astrophysical terms) is then denoted by $-\hat{\boldsymbol{z}}$. Therefore, the XY plane is appropriate for defining the Stokes parameters.

Let us assume, without loss of generality, that $\hat{\boldsymbol{e}}_1$ is in the plane formed by $\hat{\boldsymbol{e}}_0$ and $\hat{\boldsymbol{z}}$, and that it also has a positive x component ($\hat{\boldsymbol{e}}_1\cdot\hat{\boldsymbol{x}} > 0$). The geometrical configuration is illustrated in Fig. 7.1. Note that another choice for a right-handed coordinate system could have been made by changing $\hat{\boldsymbol{e}}_1$ to $-\hat{\boldsymbol{e}}_1$ and $\hat{\boldsymbol{e}}_2$ to $-\hat{\boldsymbol{e}}_2$. In such a case, $\hat{\boldsymbol{e}}_1\cdot\hat{\boldsymbol{x}} < 0$. This dual choice reflects an ever-present ambiguity of 180° in determining the azimuthal angle of $\hat{\boldsymbol{e}}_0$ from observations. In spite of this ambiguity, the geometrical configuration is completely general.

Let us call θ the colatitude angle of $\hat{\boldsymbol{e}}_0$, i.e., the angle between $\hat{\boldsymbol{e}}_0$ and $\hat{\boldsymbol{z}}$ ($\theta \in [0,\pi]$), and φ the azimuth angle of $\hat{\boldsymbol{e}}_0$, that is, the angle between the projection of $\hat{\boldsymbol{e}}_0$ on the XY plane and the X axis ($\varphi \in [0, 2\pi]$). Note that φ is the azimuthal angle of $\hat{\boldsymbol{e}}_1$ as well. The transformation between the two orthonormal systems is given by

$$(\hat{\boldsymbol{x}}\ \hat{\boldsymbol{y}}\ \hat{\boldsymbol{z}})^{\mathrm{T}} = \mathbf{B}\,(\hat{\boldsymbol{e}}_1\ \hat{\boldsymbol{e}}_2\ \hat{\boldsymbol{e}}_0)^{\mathrm{T}}, \tag{7.1}$$

where

$$\mathbf{B} \equiv \begin{pmatrix} \cos\theta\cos\varphi & -\sin\varphi & \sin\theta\cos\varphi \\ \cos\theta\sin\varphi & \cos\varphi & \sin\theta\sin\varphi \\ -\sin\theta & 0 & \cos\theta \end{pmatrix}. \tag{7.2}$$

Note that $\mathbf{B}$ is a rotation matrix, since $\mathbf{B}\mathbf{B}^{\mathrm{T}} = \mathbf{B}^{\mathrm{T}}\mathbf{B} = \mathbb{1}$ and $\det(\mathbf{B}) = 1$.

† Although the wave equation has not been written in full here, its most general formulation contains a term with that gradient which accounts for inhomogeneities in the medium. See, for example, Eq. (5) in Chapter 1 of Born and Wolf (1993).

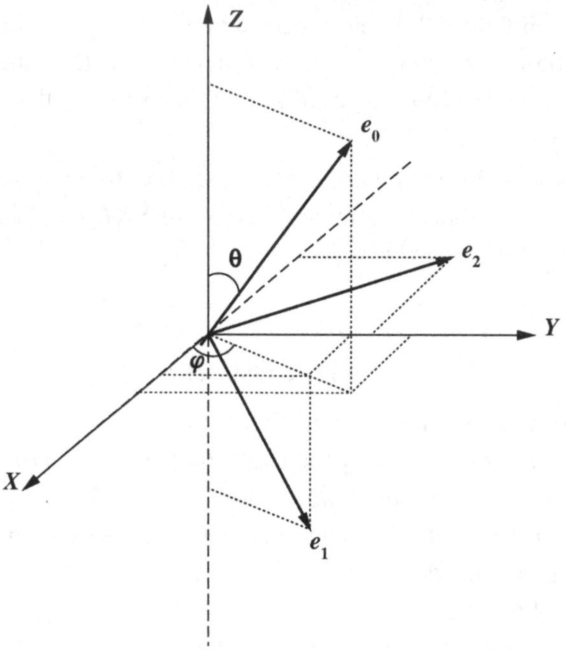

Fig. 7.1. Relevant geometry for radiative transfer.

Consider now the reference system given by the set of orthonormal complex vectors $\{\hat{\boldsymbol{e}}_{+1}, \hat{\boldsymbol{e}}_0, \hat{\boldsymbol{e}}_{-1}\}$ defined by

$$\begin{aligned} \hat{\boldsymbol{e}}_{\pm 1} &\equiv \frac{1}{\sqrt{2}}(\hat{\boldsymbol{e}}_1 \pm \mathrm{i}\hat{\boldsymbol{e}}_2), \\ \hat{\boldsymbol{e}}_0 &\equiv \hat{\boldsymbol{e}}_0. \end{aligned} \tag{7.3}$$

Of course, the norm comes from the scalar product $\boldsymbol{a} \cdot \boldsymbol{b}^*$: $\hat{\boldsymbol{e}}_\alpha \cdot \hat{\boldsymbol{e}}^*_\beta = \delta_{\alpha\beta}, \ \alpha, \beta = +1, 0, -1$. Equations (7.3) can be written in matrix form as

$$(\hat{\boldsymbol{e}}_1 \ \hat{\boldsymbol{e}}_2 \ \hat{\boldsymbol{e}}_0)^{\mathrm{T}} = \mathbf{A} \ (\hat{\boldsymbol{e}}_{+1} \ \hat{\boldsymbol{e}}_0 \ \hat{\boldsymbol{e}}_{-1})^{\mathrm{T}}, \tag{7.4}$$

where

$$\mathbf{A} \equiv \frac{1}{\sqrt{2}} \begin{pmatrix} 1 & 0 & 1 \\ -\mathrm{i} & 0 & \mathrm{i} \\ 0 & \sqrt{2} & 0 \end{pmatrix}, \tag{7.5}$$

and it can be verified that $\mathbf{A}\mathbf{A}^\dagger = \mathbf{A}^\dagger\mathbf{A} = \mathbb{1}$, whence $\mathbf{A}^{\mathrm{T}}\mathbf{A}^* = \mathbf{A}^*\mathbf{A}^{\mathrm{T}} = \mathbb{1}$, although $\det(\mathbf{A}) = -\mathrm{i}$. The index † means the adjoint, that is, the transpose of the complex conjugate.

The transformation between $\{\hat{\boldsymbol{x}}, \hat{\boldsymbol{y}}, \hat{\boldsymbol{z}}\}$ and $\{\hat{\boldsymbol{e}}_{+1}, \hat{\boldsymbol{e}}_0, \hat{\boldsymbol{e}}_{-1}\}$ is then given by

$$(\hat{\boldsymbol{x}}\ \hat{\boldsymbol{y}}\ \hat{\boldsymbol{z}})^{\mathrm{T}} = \mathbf{G}\,(\hat{\boldsymbol{e}}_{+1}\ \hat{\boldsymbol{e}}_0\ \hat{\boldsymbol{e}}_{-1})^{\mathrm{T}}, \tag{7.6}$$

where $\mathbf{G} \equiv \mathbf{BA}$ has the explicit form

$$\frac{1}{\sqrt{2}}\begin{pmatrix} \cos\theta\cos\varphi + \mathrm{i}\sin\varphi & \sqrt{2}\sin\theta\cos\varphi & \cos\theta\cos\varphi - \mathrm{i}\sin\varphi \\ \cos\theta\sin\varphi - \mathrm{i}\cos\varphi & \sqrt{2}\sin\theta\sin\varphi & \cos\theta\sin\varphi + \mathrm{i}\cos\varphi \\ -\sin\theta & \sqrt{2}\cos\theta & -\sin\theta \end{pmatrix} \tag{7.7}$$

and it can be verified that $\mathbf{G}\mathbf{G}^{\dagger} = \mathbf{G}^{\dagger}\mathbf{G} = \mathbb{1}$, whence $\mathbf{G}^{*}\mathbf{G}^{\mathrm{T}} = \mathbf{G}^{\mathrm{T}}\mathbf{G}^{*} = \mathbb{1}$, and $\det(\mathbf{G}) = -\mathrm{i}$. Therefore, matrix $\mathbf{G}^{*}$ (the inverse matrix of $\mathbf{G}^{\mathrm{T}}$) gives the transformation between the principal components of the vector electric field and its Cartesian components:

$$\begin{pmatrix} E_x(t) \\ E_y(t) \\ 0 \end{pmatrix} = \mathbf{G}^{*}\begin{pmatrix} E_{+1}(t) \\ E_0(t) \\ E_{-1}(t) \end{pmatrix}. \tag{7.8}$$

7.2 Variations of the coherency matrix along the ray path

Consider again our ray propagating along the Z axis. The relevant length along the ray path ($\hat{\boldsymbol{t}} = \hat{\boldsymbol{z}}$) is $\boldsymbol{r}\cdot\hat{\boldsymbol{z}} = z$. Hence, at a given point z in the path, the principal components of the electric field given by Eq. (6.9) can be written as

$$E_\alpha(t) = \left\{\mathcal{E}_\alpha(t)\,\mathrm{e}^{-\frac{1}{2}[\chi_\alpha - \mathrm{i}(\tilde{\chi}_\alpha + \frac{2\omega}{c})]z}\mathrm{e}^{\mathrm{i}\varpi_\alpha(t)}\right\}\mathrm{e}^{-\mathrm{i}\omega t}, \tag{7.9}$$

where the absorption and dispersion profiles are given by Eqs (6.62) and (6.63).

If we define

$$\boldsymbol{\Lambda} \equiv \frac{1}{2}\mathrm{diag}\left[\chi_{+1} - \mathrm{i}\left(\tilde{\chi}_{+1} + \frac{2\omega}{c}\right), \chi_0 - \mathrm{i}\left(\tilde{\chi}_0 + \frac{2\omega}{c}\right), \chi_{-1} - \mathrm{i}\left(\tilde{\chi}_{-1} + \frac{2\omega}{c}\right)\right] \tag{7.10}$$

then the variations of the vector electric field along the ray path at such points are given by

$$\frac{\mathrm{d}}{\mathrm{d}z}\begin{pmatrix} E_{+1} \\ E_0 \\ E_{-1} \end{pmatrix} = -\boldsymbol{\Lambda}\begin{pmatrix} E_{+1} \\ E_0 \\ E_{-1} \end{pmatrix}. \tag{7.11}$$

After transformation (7.8), Eq. (7.11) can be written as

$$\frac{\mathrm{d}}{\mathrm{d}z}\begin{pmatrix} E_x \\ E_y \\ 0 \end{pmatrix} = -\mathbf{G}^{*}\boldsymbol{\Lambda}\mathbf{G}^{\mathrm{T}}\begin{pmatrix} E_x \\ E_y \\ 0 \end{pmatrix}, \tag{7.12}$$

and, calling **T** the 2×2 matrix made from the first two rows and the first two columns of $\mathbf{G}^*\mathbf{\Lambda}\mathbf{G}^{\mathrm{T}}$, we finally obtain the variation of the Cartesian components of the electric field along the ray path:†

$$\frac{\mathrm{d}}{\mathrm{d}z}\begin{pmatrix} E_x \\ E_y \end{pmatrix} = -\mathbf{T}\begin{pmatrix} E_x \\ E_y \end{pmatrix}. \tag{7.13}$$

Since, by definition,

$$\mathbf{C} = \left\langle \begin{pmatrix} E_x \\ E_y \end{pmatrix} \begin{pmatrix} E_x^* & E_y^* \end{pmatrix} \right\rangle, \tag{7.14}$$

and because the derivative operator is linear and continuous and hence commutes with time averages, the variations of the coherency matrix along the ray path are given by

$$\frac{\mathrm{d}\mathbf{C}}{\mathrm{d}z} = -(\mathbf{TC} + \mathbf{CT}^\dagger). \tag{7.15}$$

Let $\mathbf{H} \equiv \mathbf{TC} + \mathbf{CT}^\dagger$. Then, Eq. (7.15) becomes

$$\frac{\mathrm{d}\mathbf{C}}{\mathrm{d}z} = -\mathbf{H}. \tag{7.16}$$

7.3 Variations of the Stokes parameters along the ray path

After multiplying Eq. (7.16) by the Pauli and identity matrices and taking the trace on both sides of the equation, we obtain one equation for the variation along the ray path, owing to absorption and dispersion processes, of each of the four Stokes parameters [remembering Eqs (3.16), (3.17), and (3.18)]:

$$\frac{\mathrm{d}}{\mathrm{d}z}\mathrm{Tr}\,(\mathbf{C}\boldsymbol{\sigma}_i) = -\mathrm{Tr}\,(\mathbf{H}\boldsymbol{\sigma}_i), \quad i = 0, 1, 2, 3. \tag{7.17}$$

According to Eq. (3.16), the left-hand side of Eq. (7.17) represents the derivative with respect to z of a given Stokes parameter if we neglect the dimensional constant κ that converts the Stokes parameters to specific intensity units. This neglecting of κ is justified since the constant appears on the right-hand side of the equation as well. This side can be neatly written as

$$\mathrm{Tr}\,(\mathbf{H}\boldsymbol{\sigma}_i) = \sum_{j=0}^{3} K_{ij} I_j, \tag{7.18}$$

† The column vector appearing in Eq. (7.13) is known as the Jones vector of the light beam, so that equation accounts for the variations of the Jones vector owing to absorption and dispersion processes.

where

$$K_{ij} \equiv \frac{1}{2} \mathrm{Tr}\,(\mathbf{T}\boldsymbol{\sigma}_j \boldsymbol{\sigma}_i + \boldsymbol{\sigma}_j \mathbf{T}^{\mathrm{T}} \boldsymbol{\sigma}_i) \tag{7.19}$$

and $I_0 = I$, $I_1 = Q$, $I_2 = U$, and $I_3 = V$. Equations (7.18) and (7.19) come directly from applying Eq. (3.18) to the definition of matrix **H**.

Written in full, and neglecting κ, the right-hand side of Eq. (7.17) reads

$$\begin{aligned}
\mathrm{Tr}\,(\mathbf{H}\boldsymbol{\sigma}_0) &= \mathrm{Re}\,(T_{11}+T_{22})I \quad + \mathrm{Re}\,(T_{11}-T_{22})Q \\
&\quad + \mathrm{Re}\,(T_{12}+T_{21})U + \mathrm{Im}\,(T_{12}-T_{21})V \\
\mathrm{Tr}\,(\mathbf{H}\boldsymbol{\sigma}_1) &= \mathrm{Re}\,(T_{11}-T_{22})I \quad + \mathrm{Re}\,(T_{11}+T_{22})Q \\
&\quad + \mathrm{Re}\,(T_{12}-T_{21})U + \mathrm{Im}\,(T_{12}+T_{21})V \\
\mathrm{Tr}\,(\mathbf{H}\boldsymbol{\sigma}_2) &= \mathrm{Re}\,(T_{12}+T_{21})I \quad - \mathrm{Re}\,(T_{12}-T_{21})Q \\
&\quad + \mathrm{Re}\,(T_{11}+T_{22})U - \mathrm{Im}\,(T_{11}-T_{22})V \\
\mathrm{Tr}\,(\mathbf{H}\boldsymbol{\sigma}_3) &= \mathrm{Im}\,(T_{12}-T_{21})I \quad - \mathrm{Im}\,(T_{12}+T_{21})Q \\
&\quad + \mathrm{Im}\,(T_{11}-T_{22})U + \mathrm{Re}\,(T_{11}+T_{22})V.
\end{aligned} \tag{7.20}$$

In summary, it is easy to see that Eq. (7.17) represents an equation for the Stokes vector of the form

$$\frac{\mathrm{d}}{\mathrm{d}z}\begin{pmatrix} I \\ Q \\ U \\ V \end{pmatrix} = -\begin{pmatrix} \eta_I & \eta_Q & \eta_U & \eta_V \\ \eta_Q & \eta_I & \rho_V & -\rho_U \\ \eta_U & -\rho_V & \eta_I & \rho_Q \\ \eta_V & \rho_U & -\rho_Q & \eta_I \end{pmatrix}\begin{pmatrix} I \\ Q \\ U \\ V \end{pmatrix}, \tag{7.21}$$

where the different matrix elements on the right-hand side are easily identifiable from Eq. (7.20):

$$\begin{aligned}
\eta_I &\equiv \mathrm{Re}\,(T_{11}+T_{22}), & \eta_Q &\equiv \mathrm{Re}\,(T_{11}-T_{22}), \\
\eta_U &\equiv \mathrm{Re}\,(T_{12}+T_{21}), & \eta_V &\equiv \mathrm{Im}\,(T_{12}-T_{21}), \\
\rho_Q &\equiv -\mathrm{Im}\,(T_{11}-T_{22}), & \rho_U &\equiv -\mathrm{Im}\,(T_{12}+T_{21}), \\
\rho_V &\equiv \mathrm{Re}\,(T_{12}-T_{21}).
\end{aligned} \tag{7.22}$$

By definition of matrix $\mathbf{T}$ we have

$$
\begin{aligned}
T_{11} + T_{22} &= \Lambda_0 \sin^2\theta + \frac{1}{2}(\Lambda_{+1} + \Lambda_{-1})(1 + \cos^2\theta), \\
T_{12} + T_{21} &= \left[\Lambda_0 - \frac{1}{2}(\Lambda_{+1} + \Lambda_{-1})\right] \sin^2\theta \sin 2\varphi, \\
T_{11} - T_{22} &= \left[\Lambda_0 - \frac{1}{2}(\Lambda_{+1} + \Lambda_{-1})\right] \sin^2\theta \cos 2\varphi, \\
T_{12} - T_{21} &= \mathrm{i}(\Lambda_{-1} - \Lambda_{+1}) \cos\theta,
\end{aligned}
\tag{7.23}
$$

where $\Lambda_\alpha,\ \alpha = +1, 0, -1$, are the diagonal elements of matrix $\mathbf{\Lambda}$ [Eq. (7.10)]. Notably, $T_{12} - T_{21}$, the difference involved in the definition of η_V and ρ_V, is proportional to the difference of *right-handed minus left-handed* terms. Taking real and imaginary parts of the above expressions one finally gets

$$
\begin{aligned}
\eta_I &= \frac{1}{2}\left\{\chi_0 \sin^2\theta + \frac{1}{2}\left[\chi_{+1} + \chi_{-1}\right](1 + \cos^2\theta)\right\}, \\
\eta_Q &= \frac{1}{2}\left\{\chi_0 - \frac{1}{2}\left[\chi_{+1} + \chi_{-1}\right]\right\} \sin^2\theta \cos 2\varphi, \\
\eta_U &= \frac{1}{2}\left\{\chi_0 - \frac{1}{2}\left[\chi_{+1} + \chi_{-1}\right]\right\} \sin^2\theta \sin 2\varphi, \\
\eta_V &= \frac{1}{2}\left[\chi_{-1} - \chi_{+1}\right] \cos\theta,
\end{aligned}
\tag{7.24}
$$

and

$$
\begin{aligned}
\rho_Q &= \frac{1}{2}\left\{\tilde{\chi}_0 - \frac{1}{2}\left[\tilde{\chi}_{+1} + \tilde{\chi}_{-1}\right]\right\} \sin^2\theta \cos 2\varphi, \\
\rho_U &= \frac{1}{2}\left\{\tilde{\chi}_0 - \frac{1}{2}\left[\tilde{\chi}_{+1} + \tilde{\chi}_{-1}\right]\right\} \sin^2\theta \sin 2\varphi, \\
\rho_V &= \frac{1}{2}\left[\tilde{\chi}_{-1} - \tilde{\chi}_{+1}\right] \cos\theta.
\end{aligned}
\tag{7.25}
$$

Therefore, *the matrix elements of Eq. (7.21) are made up of the various absorption and dispersion profiles characteristic of the medium and all the geometry relevant to the problem.* That equation can be written in a more compact matrix form as

$$
\frac{\mathrm{d}\boldsymbol{I}}{\mathrm{d}z} = -\mathbf{K}\boldsymbol{I}, \tag{7.26}
$$

where **K**, anticipated in Eq. (7.19), may be called the *propagation matrix*. Some texts and papers use the less fortunate names of "absorption" or "absorption–dispersion" matrix. These names come from an extrapolation of the *scalar* transfer equation for isotropic media (also called "for unpolarized light"), which we shall discuss later. The term *propagation matrix* seems more suitable because not only are absorption and dispersion described by the matrix but also birefringence (or dichroism). In fact, the symmetries of matrix **K** allow us to decompose it in three matrices:

$$\mathbf{K} = \begin{pmatrix} \eta_I & 0 & 0 & 0 \\ 0 & \eta_I & 0 & 0 \\ 0 & 0 & \eta_I & 0 \\ 0 & 0 & 0 & \eta_I \end{pmatrix} + \begin{pmatrix} 0 & \eta_Q & \eta_U & \eta_V \\ \eta_Q & 0 & 0 & 0 \\ \eta_U & 0 & 0 & 0 \\ \eta_V & 0 & 0 & 0 \end{pmatrix} + \begin{pmatrix} 0 & 0 & 0 & 0 \\ 0 & 0 & \rho_V & -\rho_U \\ 0 & -\rho_V & 0 & \rho_Q \\ 0 & \rho_U & -\rho_Q & 0 \end{pmatrix}. \tag{7.27}$$

In Eq. (7.27), the first, diagonal matrix corresponds to absorption phenomena. Energy from all polarization states is withdrawn by the medium: all four Stokes parameters evolve the same way. The second, symmetric matrix corresponds to dichroism: some polarized components of the beam are extinguished more than others because the matrix elements are generally different. Finally, the third, antisymmetric matrix corresponds to dispersion: phase shifts that take place during the propagation to some extent change different states of linear polarization among themselves (*Faraday rotation*) and states of linear polarization with states of circular polarization (*Faraday pulsation*).

7.4 Some properties of the propagation matrix

Although Eq. (7.26) does not represent a finite transformation between two sets of Stokes parameters like Eq. (4.28), the role of the propagation matrix is reminiscent of that of the Mueller matrix. We could call it a pseudo-Mueller matrix or an "infinitesimal" Mueller matrix. With this interpretation, the evolution of the Stokes vector consists in the incoherent (infinitesimal) subtraction of light beams from the original one. It is by no means strange, then, that matrix **K** displays properties already known for Mueller matrices. In particular, the first element, η_I, is non-negative [Eq. (7.24)] and the elements of the first line and of the first column verify that

$$\eta_I^2 \geq \eta_Q^2 + \eta_U^2 + \eta_V^2, \tag{7.28}$$

equivalent to Eqs (4.33) and (4.34). This result comes directly from the expressions for those elements in terms of the absorption and dispersion profiles. In fact, after some algebra, Eqs (7.24) give

$$\eta_I^2 - \eta_Q^2 - \eta_U^2 - \eta_V^2 = \chi_{+1}\chi_{-1}\cos^2\theta + \frac{1}{2}\chi_0(\chi_{+1} + \chi_{-1})\sin^2\theta, \tag{7.29}$$

and the right-hand side of this equation is always non-negative. As is usual for Mueller matrices, these four matrix elements constitute the diattenuation and the polarizance vectors so they are responsible for both the variations of the total intensity of the light beam as a function of the state of polarization and of the empolarizing capabilities of the medium (i.e., of its capability to increase the degree of polarization). Should the last three elements of the first column be zero, unpolarized light would remain unpolarized throughout the medium. In fact, we have already noted that the leftmost matrix on the right-hand side of Eq. (7.27) is unable to alter the degree of polarization of light.

The only way that $\eta_Q = \eta_U = \eta_V = 0$ is for the three absorption profiles χ_α to be the same, i.e., for absorption to be isotropic. But if absorption is isotropic so is dispersion because Eqs (6.30) and (6.32) tell us that $\tilde{\chi}_\alpha = (\chi_\alpha/\Gamma_\alpha)\,(\nu_{0,\alpha} - \nu)$. This is the case, for instance, for continuum radiation, which we have discussed in Section 6.3. Depending on the geometry of the problem, some η's (and the corresponding ρ's) may be zero. Such particular cases will be discussed later.

Much as I, V, and $\sqrt{Q^2 + U^2}$ are intrinsic parameters of the light beam regardless of the reference frame (see Section 4.6.1), Eqs (7.24) and (7.25) readily show that

$$\eta_L^2 \equiv \eta_Q^2 + \eta_U^2$$

and

$$\rho_L^2 \equiv \rho_Q^2 + \rho_U^2$$

are independent of the reference system used to define the Stokes parameters, since they do not depend on the azimuth angle, φ. As a matter of fact, a rotation of the reference system for measuring the Stokes parameters (the XY plane) by an angle γ (see Section 4.6.1) transforms the propagation matrix into

$$\begin{pmatrix} \eta_I & \eta'_Q & \eta'_U & \eta_V \\ \eta'_Q & \eta_I & \rho_V & -\rho'_U \\ \eta'_U & -\rho_V & \eta_I & \rho'_Q \\ \eta_V & \rho'_u & -\rho'_Q & \eta_I \end{pmatrix},$$

where it is evident that η_I, η_V, and ρ_V are invariant against the transformation. In contrast, η_Q, η_U, ρ_Q and ρ_U are modified as

$$\begin{aligned}\eta'_Q &= c_2\eta_Q - s_2\eta_U,\\ \eta'_U &= s_2\eta_Q + c_2\eta_U,\\ \rho'_Q &= c_2\rho_Q - s_2\rho_U,\\ \rho'_Q &= s_2\rho_Q + c_2\rho_U,\end{aligned}$$

whence $\eta'^2_Q + \eta'^2_U = \eta^2_Q + \eta^2_U$ and $\rho'^2_Q + \rho'^2_U = \rho^2_Q + \rho^2_U$. Here, $c_2 = \cos 2\gamma$ and $s_2 = \sin 2\gamma$.

It is also very interesting to understand the evolution of the degree of polarization of light traveling through an absorptive and dispersive medium. Let us consider the derivative of the quantity $f \equiv I^2 - Q^2 - U^2 - V^2 = I^2(1 - p^2)$. By writing

$$f = \boldsymbol{I}^{\mathrm{T}}\mathbf{M}\boldsymbol{I},$$

where $\mathbf{M}$ is the Minkowsky metric matrix

$$\mathbf{M} \equiv \begin{pmatrix} 1 & 0 & 0 & 0 \\ 0 & -1 & 0 & 0 \\ 0 & 0 & -1 & 0 \\ 0 & 0 & 0 & -1 \end{pmatrix},$$

it is easy to check that

$$\frac{\mathrm{d}f}{\mathrm{d}z} = -\boldsymbol{I}^{\mathrm{T}}\left[(\mathbf{MK})^{\mathrm{T}} + \mathbf{MK}\right]\boldsymbol{I} = -2\eta_I f.$$

The solution of the above equation is

$$f(z) = f(z_0)\,\mathrm{e}^{-2\int_{z_0}^{z}\eta_I(z')\,\mathrm{d}z'},$$

that is, f decreases exponentially from its initial value. Since f and η_I are always non-negative, the derivative must always be non-positive. Hence, the square degree of polarization turns out to be

$$p^2(z) = 1 - \frac{f(z_0)}{I^2(z)}\,\mathrm{e}^{-2\int_{z_0}^{z}\eta_I(z')\,\mathrm{d}z'}. \tag{7.30}$$

Since $0 \leq p^2 \leq 1$ at all points, the second term on the right of Eq. (7.30) is also bounded between 0 and 1, which means that the ratio $I^2(z_0)\left[1 - p^2(z_0)\right]/I^2(z)$ always increases more slowly than exponentially. It is noteworthy that no matter what the initial degree of polarization, a medium with η_I = constant transmits a beam that is asymptotically completely polarized. Moreover, if the initial degree of polarization is unity, the light stays totally polarized throughout the medium whatever η_I.

A medium which is unable to decrease the degree of polarization of totally polarized light is said to be *non-depolarizing*. Hence, we have deduced that absorption and dispersion processes are non-depolarizing; in other words, a propagation matrix like **K** is a sufficient condition for the medium to be non-depolarizing. Landi Degl'Innocenti and Landi Degl'Innocenti (1981) showed that every non-depolarizing medium must necessarily have a propagation matrix with the symmetries of matrix **K**. Therefore, we have a necessary and sufficient condition and can then conclude that *depolarizing effects can appear in the transport of radiative energy only through the emission terms that we have so far neglected.*

7.5 Emission processes

Equation (7.26) accounts only for absorption and dispersion effects. To understand fully the variations of the Stokes vector along the ray path, we need a new term including the emissive properties of the medium. Such a term is customarily represented by a Stokes vector,

$$\boldsymbol{j} \equiv \begin{pmatrix} j_I \\ j_Q \\ j_U \\ j_V \end{pmatrix}, \tag{7.31}$$

whose characteristics are to be determined in this section. Therefore, the *radiative transfer equation* (RTE) will read

$$\frac{\mathrm{d}\boldsymbol{I}}{\mathrm{d}z} = -\mathbf{K}\boldsymbol{I} + \boldsymbol{j}, \tag{7.32}$$

hence including the (incoherent) subtraction (first term) and the (incoherent) addition of infinitesimal Stokes beams. Note that the RTE can be considered as a single differential equation for vector quantities, or as a system of four coupled differential equations for scalar quantities: no solution can be found for a given Stokes parameter without solving the entire system.

Although customary in most textbooks, it is very difficult to establish conceptually a neat distinction between "true" absorption on the one hand and emission and scattering processes on the other. Many ambiguities arise when one scrutinizes the many microscopic phenomena giving rise to the interaction between radiation and matter. We refer the interested reader to the excellent didactic discussions of Cox and Giuli (1968) and Mihalas (1978). We shall not enter into details but a few words are necessary to establish the framework (i.e. the assumptions) within which our developments are valid. In fact, the phenomenological description of absorption on the basis of the Lorentz electron model may include processes typically characterized as scattering.

In a broad sense, one could say that "true" absorption processes include those supplying energy to the thermal pool of matter at the expense of electromagnetic energy, while scattering influences the energetic balance by deviating photons from their original path after having collided with material particles; in such processes, not only the direction of light may have changed but also the frequency of photons, although only a slight influence (if any) is effected on the kinetic energy of matter. This broad difference clearly illustrates that some phenomena are mostly dependent upon the *local* value of the material thermodynamic variables (true absorption and emission), while others (scattering) depend mainly on the radiation field, which may therefore be connecting fairly distant places of the medium. Hence, the general radiative transfer problem has a marked *non-linear* character: one is seeking a transfer equation that explains the evolution of the Stokes vector of light once the material properties are known; however, the state of matter is dependent upon the characteristics of the radiation field itself, and the formulation of statistical equilibrium equations is required. The simultaneous solution of both the radiative transfer equation and the statistical equilibrium equations is mandatory.

This book deliberately skips this general problem of polarized radiative transfer and tries to provide but a first step in our understanding of this enormous and exciting puzzle. The discussion will be limited to the simplest case of those situations for which departures from the so-called *local thermodynamic equilibrium* (LTE) approximation are negligible. More general, rigorous treatments can be found in the monographs by Stenflo (1994) and Landi Degl'Innocenti and Landolfi (2003). Although manifestly inconsistent from a conceptual point of view, the LTE approximation turns out to be a satisfactory approach for many applications of interest in both the laboratory and the astrophysical contexts that will be discussed in the following chapters. In particular, the formation of a large number of spectral lines in the photosphere of the Sun and other cool stars is adequately described by assuming that radiative transfer has taken place in conditions of local thermodynamic equilibrium.

The LTE hypothesis basically consists in assuming that only radiation (and not matter) is allowed to deviate from a thermodynamic equilibrium situation because of the transport. All the thermodynamic properties of matter are assumed to be governed by the *thermodynamic equilibrium* equations but at the *local* values of temperature, T, and density, ρ. Hence, the local distribution of velocities is Maxwellian [Eq. (6.41)]; the local number of absorbers and emitters in the various quantum states is given by the Boltzmann and Saha equations, and Kirchhoff's law is verified:

$$\boldsymbol{j}^{\mathrm{T}} = B_\nu(T)\,(\eta_I, \eta_Q, \eta_U, \eta_V), \tag{7.33}$$

thus providing the needed emission term having exactly the same frequency shape as the absorption profiles. Naturally, the emission vector includes only the first line of the propagation matrix since it is these elements that affect the modifications of the total intensity (first Stokes parameter) of the light beam. Inequality (7.28) ensures the physical meaning of $\boldsymbol{j}$ as a Stokes vector. Moreover, Eq. (7.29) implies that the emission vector is partially polarized, and that it is then responsible for depolarization effects along the ray path, according to the discussions in the preceding section and Section 3.4.

The physical conditions in a stellar atmosphere are distant from those of an adiabatically isolated enclosure in which thermodynamic equilibrium prevails. Fairly large gradients exist in the temperature and density throughout, and radiation cannot be isotropic since light can escape from the outermost parts of the stellar envelope. Nonetheless, one can conceive of some mechanism that enforces LTE conditions, such as frequent collisions between material particles, which somehow decouple matter from radiation. We may then expect that the larger the density, the closer the conditions in the medium to LTE. Hence, departures from this approximation can be foreseen for lines formed mostly in the outermost stellar photosphere and in the chromosphere, where densities are not high enough.

Pure scattering processes may exhibit clear differences between the absorption and emission profiles so that they may avoid the use of the Kirchhoff's relationship (7.33). This is not the case, however, for thermodynamic equilibrium when *natural excitation* prevails, when absorption and emission profiles are the same. Nor is it the case for those conditions that make valid the *complete redistribution* approximation. According to this approximation there is no correlation between the frequencies of the incoming and scattered photons because, for example, there is a continuous supply of atoms in the upper state of the transition due, for example, to collisions. Again, an adequate rate of collisions helps in keeping the medium such that absorption and emission profiles are related by the Planck function at the local temperature.

In summary, we may say that within the LTE and complete redistribution approximations, light streams through a plane–parallel or stratified medium according to the following RTE:

$$\frac{\mathrm{d}}{\mathrm{d}z}\begin{pmatrix} I \\ Q \\ U \\ V \end{pmatrix} = -\begin{pmatrix} \eta_I & \eta_Q & \eta_U & \eta_V \\ \eta_Q & \eta_I & \rho_V & -\rho_U \\ \eta_U & -\rho_V & \eta_I & \rho_Q \\ \eta_V & \rho_U & -\rho_Q & \eta_V \end{pmatrix}\begin{pmatrix} I \\ Q \\ U \\ V \end{pmatrix} + B_\nu(T)\begin{pmatrix} \eta_I \\ \eta_Q \\ \eta_U \\ \eta_V \end{pmatrix} \tag{7.34}$$

or, more compactly,

$$\frac{\mathrm{d}}{\mathrm{d}z}\begin{pmatrix} I \\ Q \\ U \\ V \end{pmatrix} = -\begin{pmatrix} \eta_I & \eta_Q & \eta_U & \eta_V \\ \eta_Q & \eta_I & \rho_V & -\rho_U \\ \eta_U & -\rho_V & \eta_I & \rho_Q \\ \eta_V & \rho_U & -\rho_Q & \eta_V \end{pmatrix}\begin{pmatrix} I - B_\nu(T) \\ Q \\ U \\ V \end{pmatrix}, \tag{7.35}$$

where

$$\boldsymbol{S} \equiv (B_\nu(T), 0, 0, 0)^{\mathrm{T}} \tag{7.36}$$

is called the *source function* vector.

7.6 The RTE for spectral line formation

Yet some manipulations remain in order to shape the RTE into a common and useful form for describing the formation of spectral lines in an anisotropic medium. Contributions from both continuum-forming processes and pure line-forming processes ought to be added in the propagation matrix, so that $\mathbf{K} = \mathbf{K}_{\mathrm{cont}} + \mathbf{K}_{\mathrm{lin}}$.

According to the discussions of Section 7.4, the propagation matrix for those processes giving rise to the continuum is proportional to the 4×4 identity matrix,

$$\mathbf{K}_{\mathrm{cont}} = \chi_{\mathrm{cont}} \mathbb{1}, \tag{7.37}$$

where χ_{cont} is the frequency-independent absorption coefficient for the continuum (see Section 6.3). On the other hand, according to Eqs (7.24), (7.25), (6.62), and (6.63), we can write

$$\mathbf{K}_{\mathrm{lin}} = \frac{N\pi \mathrm{e}_0^2 f}{mc} \boldsymbol{\Phi}, \tag{7.38}$$

where $\boldsymbol{\Phi}$ contains only the normalized absorption and dispersion profiles $\phi_\alpha(u_{0,\alpha}, a_\alpha)$ and $\psi_\alpha(u_{0,\alpha}, a_\alpha)$. Therefore, the total propagation matrix turns out to be

$$\mathbf{K} = \chi_{\mathrm{cont}}(\mathbb{1} + \eta_0 \boldsymbol{\Phi}), \tag{7.39}$$

where, by definition, η_0 is the *line-to-continuum absorption coefficient ratio*,

$$\eta_0 \equiv \frac{\chi_{\mathrm{lin}}}{\chi_{\mathrm{cont}}} = \frac{N\pi \mathrm{e}_0^2 f}{mc\chi_{\mathrm{cont}}}. \tag{7.40}$$

Let us introduce now the *continuum optical depth* variable via the expression

$$\tau_{\mathrm{c}} \equiv \int_z^{z_0} \chi_{\mathrm{cont}}\, \mathrm{d}z. \tag{7.41}$$

Note that the above definition implies that optical depths are measured along the ray path but in the opposite direction (i.e., $-\hat{z}$) and the origin ($\tau_{\mathrm{c}} = 0$) is located in the outermost boundary of the medium (z_0), where the observer is located. If, according to Eq. (6.16), we write the integrand of Eq. (7.41) as $1/\ell_{\rm cont}$, the interpretation of the optical depth is natural: $\tau_{\rm c}$ represents the (dimensionless) number of mean free paths of continuum photons between the outermost boundary and point z.

Using $\tau_{\rm c}$ as the independent variable, the RTE can be written as

$$\frac{\mathrm{d}\boldsymbol{I}}{\mathrm{d}\tau_{\rm c}} = \mathbf{K}\,(\boldsymbol{I} - \boldsymbol{S}), \tag{7.42}$$

where we have kept the symbol $\mathbf{K}$ for the propagation matrix; that is, from now on

$$\mathbf{K} = \mathbb{1} + \eta_0 \boldsymbol{\Phi}. \tag{7.43}$$

Thus, the matrix elements of $\mathbf{K}$ ought to be recast in the following form:

$$\begin{aligned}
\eta_I &= 1 + \frac{\eta_0}{2}\left\{\phi_0 \sin^2\theta + \frac{1}{2}\left[\phi_{+1} + \phi_{-1}\right](1+\cos^2\theta)\right\}, \\
\eta_Q &= \frac{\eta_0}{2}\left\{\phi_0 - \frac{1}{2}\left[\phi_{+1} + \phi_{-1}\right]\right\}\sin^2\theta\cos 2\varphi, \\
\eta_U &= \frac{\eta_0}{2}\left\{\phi_0 - \frac{1}{2}\left[\phi_{+1} + \phi_{-1}\right]\right\}\sin^2\theta\sin 2\varphi, \\
\eta_V &= \frac{\eta_0}{2}\left[\phi_{-1} - \phi_{+1}\right]\cos\theta,
\end{aligned} \tag{7.44}$$

and

$$\begin{aligned}
\rho_Q &= \frac{\eta_0}{2}\left\{\psi_0 - \frac{1}{2}\left[\psi_{+1} + \psi_{-1}\right]\right\}\sin^2\theta\cos 2\varphi, \\
\rho_U &= \frac{\eta_0}{2}\left\{\psi_0 - \frac{1}{2}\left[\psi_{+1} + \psi_{-1}\right]\right\}\sin^2\theta\sin 2\varphi, \\
\rho_V &= \frac{\eta_0}{2}\left[\psi_{-1} - \psi_{+1}\right]\cos\theta.
\end{aligned} \tag{7.45}$$

7.7 Radiative transfer through isotropic media

Let us here consider a particularly interesting case. Let us assume that the medium is isotropic. Then, according to the discussions of Sections 6.2 and 7.4, we have $\eta_Q = \eta_U = \eta_V = \rho_Q = \rho_U = \rho_V = 0$, because $\phi_\alpha(u_{0,\alpha}, a_\alpha) = \phi(u_0, a)$, $\forall\,\alpha = +1, 0, -1$ and $\psi_\alpha(u_{0,\alpha}, a_\alpha) = \psi(u_0, a)$, $\forall\,\alpha = +1, 0, -1$. The propagation matrix turns out to be diagonal. This enormously simplifies the problem since the RTE becomes four *uncoupled* equations each for a given Stokes parameter:

$$\frac{\mathrm{d}I}{\mathrm{d}\tau_{\mathrm{c}}} = (1 + \eta_0 \phi)\,(I - B_\nu), \tag{7.46}$$

$$\frac{\mathrm{d}Q}{\mathrm{d}\tau_{\mathrm{c}}} = (1 + \eta_0 \phi)\,Q, \tag{7.47}$$

$$\frac{\mathrm{d}U}{\mathrm{d}\tau_{\mathrm{c}}} = (1 + \eta_0 \phi)\,U, \tag{7.48}$$

and

$$\frac{\mathrm{d}V}{\mathrm{d}\tau_{\mathrm{c}}} = (1 + \eta_0 \phi)\,V. \tag{7.49}$$

Hence, one can solve for each Stokes parameter independently of the others. Equation (7.46) is usually called the *radiative transfer equation for unpolarized light* but here we can easily appreciate the unsuitability of such a name. This equation is one among the four that result for the particular case of an isotropic medium. Only when the boundary condition (obviously required for solving the equations) assumes that light is originally unpolarized ($Q_0 = U_0 = V_0 = 0$) will Eqs (7.47), (7.48), and (7.49) predict that light will remain unpolarized throughout the medium. No Q, U, or V can be generated from zero because of the absence of a source term in their equations. This assumption is often employed for spectral line formation in stellar photospheres and hence the name. In any other case (e.g., the transfer of an originally polarized beam through an isotropic slab), all four equations must be solved to understand the polarization state of the light beam.

7.8 Propagation along the optical axis and in a perpendicular direction

This section is devoted to two particular cases for which the RTE becomes significantly simplified like that when light propagates through an isotropic medium.

Imagine that the optical axis, $\hat{\boldsymbol{e}}_0$, stays unaltered throughout the medium, and that the ray path coincides with it (or with $-\hat{\boldsymbol{e}}_0$). In such a case, the colatitude angle $\theta = 0\ (\pi)$ so that the matrix elements of $\mathbf{K}$ become

$$\eta_I = 1 + \frac{\eta_0}{2}\left[\phi_{+1} + \phi_{-1}\right], \tag{7.50}$$

$$\eta_Q = \eta_U = \rho_Q = \rho_U = 0, \tag{7.51}$$

$$\eta_V = \frac{\eta_0}{2}\left[\phi_{-1} - \phi_{+1}\right], \tag{7.52}$$

and

$$\rho_V = \frac{\eta_0}{2}\left[\psi_{-1} - \psi_{+1}\right]. \tag{7.53}$$

When $\theta = \pi$, η_V and ρ_V become $-\eta_V$ and $-\rho_V$.

We can readily understand why ϕ_0 and ψ_0 have disappeared: light polarized along $\hat{\boldsymbol{e}}_0$ cannot propagate along $\hat{\boldsymbol{e}}_0$ because electromagnetic waves are transversal.

The transfer equation adopts the simpler shape

$$\frac{\mathrm{d}}{\mathrm{d}\tau_\mathrm{c}}\begin{pmatrix} I \\ Q \\ U \\ V \end{pmatrix} = \begin{pmatrix} \eta_I & 0 & 0 & \eta_V \\ 0 & \eta_I & \rho_V & 0 \\ 0 & -\rho_V & \eta_I & 0 \\ \eta_V & 0 & 0 & \eta_I \end{pmatrix}\begin{pmatrix} I - B_\nu \\ Q \\ U \\ V \end{pmatrix}, \tag{7.54}$$

where only Stokes I and V, on the one hand, and Stokes Q and U, on the other, are coupled. Since no source term appears in the Stokes Q and U equations, we may conclude that, if light is originally unpolarized, no linear polarization can be produced as a consequence of the transfer. On the contrary, two independent scalar equations can be obtained for $I + V$ and for $I - V$, namely,

$$\frac{\mathrm{d}}{\mathrm{d}\tau_\mathrm{c}}(I \pm V) = (\eta_I \pm \eta_V)\,(I \pm V - B_\nu), \tag{7.55}$$

which are formally identical to that of Stokes I for the isotropic case.

In summary, we can say that if light is originally unpolarized before entering the medium, only circular polarization can be produced when the ray path coincides with the optical axis of the medium.

Imagine now that $\hat{\boldsymbol{t}} \cdot \hat{\boldsymbol{e}}_0 = 0$, so that $\theta = \pi/2$, and that φ is fixed throughout the medium. The matrix elements reduce to

$$\eta_I = 1 + \frac{\eta_0}{2}\left\{\phi_0 + \frac{1}{2}\left[\phi_{+1} + \phi_{-1}\right]\right\}, \tag{7.56}$$

$$\eta_Q = \frac{\eta_0}{2}\left\{\phi_0 - \frac{1}{2}\left[\phi_{+1} + \phi_{-1}\right]\right\}\cos 2\varphi, \tag{7.57}$$

$$\eta_U = \frac{\eta_0}{2}\left\{\phi_0 - \frac{1}{2}\left[\phi_{+1} + \phi_{-1}\right]\right\}\sin 2\varphi, \tag{7.58}$$

$$\rho_Q = \frac{\eta_0}{2}\left\{\psi_0 - \frac{1}{2}\left[\psi_{+1} + \psi_{-1}\right]\right\}\cos 2\varphi, \tag{7.59}$$

$$\rho_U = \frac{\eta_0}{2}\left\{\psi_0 - \frac{1}{2}\left[\psi_{+1} + \psi_{-1}\right]\right\} \sin 2\varphi, \tag{7.60}$$

$$\eta_V = \rho_V = 0, \tag{7.61}$$

and the RTE to

$$\frac{\mathrm{d}}{\mathrm{d}\tau_c}\begin{pmatrix} I \\ Q \\ U \\ V \end{pmatrix} = \begin{pmatrix} \eta_I & \eta_Q & \eta_U & 0 \\ \eta_Q & \eta_I & 0 & -\rho_U \\ \eta_U & 0 & \eta_I & \rho_Q \\ 0 & \rho_U & -\rho_Q & \eta_I \end{pmatrix} \begin{pmatrix} I - B_\nu \\ Q \\ U \\ V \end{pmatrix}. \tag{7.62}$$

Let us rotate the reference system for measuring the Stokes parameters by an angle $-\varphi$, opposite of the current azimuth angle. According to the discussion of Section 7.4, η_I, η_V, and ρ_V remain the same whilst

$$\eta'_Q = \frac{\eta_0}{2}\left\{\phi_0 - \frac{1}{2}\left[\phi_{+1} + \phi_{-1}\right]\right\} = \sqrt{\eta_Q^2 + \eta_U^2},$$

$$\rho'_Q = \frac{\eta_0}{2}\left\{\psi_0 - \frac{1}{2}\left[\psi_{+1} + \psi_{-1}\right]\right\} = \sqrt{\rho_Q^2 + \rho_U^2},$$

and

$$\eta'_U = \rho'_U = 0.$$

In this new reference system, the RTE reads

$$\frac{\mathrm{d}}{\mathrm{d}\tau_c}\begin{pmatrix} I \\ Q' \\ U' \\ V \end{pmatrix} = \begin{pmatrix} \eta_I & \eta'_Q & 0 & 0 \\ \eta'_Q & \eta_I & 0 & 0 \\ 0 & 0 & \eta_I & \rho'_Q \\ 0 & 0 & -\rho'_Q & \eta_I \end{pmatrix} \begin{pmatrix} I - B_\nu \\ Q' \\ U' \\ V \end{pmatrix}. \tag{7.63}$$

Note that only Stokes Q and U get transformed whilst I and V are invariant.

We have succeeded in decoupling Stokes I and Q' on the one hand, and Stokes U' and V on the other. Since the latter have no source terms in their equations, light on input with $U'_0 = V_0 = 0$ will remain throughout the medium with $U' = V = 0$. This is not always the case, however. If U'_0 and V_0 are different from zero, a "transfer" of Stokes V to U', and vice versa, takes place as long as light travels through the medium. With the equations for Stokes I and Q' a result is found similar to that of the propagation along the optical axis, but this time polarization can only be linear if the original light is unpolarized:

$$\frac{\mathrm{d}}{\mathrm{d}\tau_c}(I \pm Q') = (\eta_I \pm \eta'_Q)\,(I \pm Q' - B_\nu). \tag{7.64}$$

Recommended bibliography

Born, M. and Wolf, E. (1993). *Principles of optics*, 6th edition (Pergamon Press: Oxford). Chapter 10. Sections 10.2–10.3, 10.7.3, and 10.8.

Chandrasekhar, S. (1946). On the radiative equilibrium of a stellar atmosphere. X. *Astrophys. J.* **103**, 351.

Chandrasekhar, S. (1946). On the radiative equilibrium of a stellar atmosphere. XI. *Astrophys. J.* **104**, 110.

Chandrasekhar, S. (1947). On the radiative equilibrium of a stellar atmosphere. XV. *Astrophys. J.* **105**, 424.

Chandrasekhar, S. (1960). *Radiative transfer* (Dover Publications, Inc.: New York). Chapter 1.

Cox, J.P. and Giuli, R.T. (1968). *Principles of stellar structure. Vol. I: Physical Principles* (Gordon and Breach, Science Publishers: New York). Chapter 2, Section 2.8.

Jefferies, J., Lites, B.W., and Skumanich, A.P. (1989). Transfer of line radiation in a magnetic field. *Astrophys. J.* **343**, 920.

Landi Degl'Innocenti, E. (1992). Magnetic field measurements, in *Solar observations: techniques and interpretations*. F. Sánchez, M. Collados, and M. Vázquez (eds.) (Cambridge University Press: Cambridge). Sections 4–8.

Landi Degl'Innocenti, E. and Landi Degl'Innocenti, M. (1981). Radiative transfer for polarized radiation: symmetry properties and geometrical interpretation. *Il Nuovo Cimento* **62 B**, 1.

Landi Degl'Innocenti, E. and Landolfi, M. (2003). *Polarization in spectral lines* (Kluwer Academic Publishers: Dordrecht), in the press.

Mihalas, D. (1978). *Stellar atmospheres* (W.H. Freeman and Company: San Francisco). Chapter 2, Section 2.1.

Rachkovsky, D.N. (1962). Magneto-optical effects in spectral lines of sunspots (in Russian). *Izv. Krymsk. Astrofiz. Obs.* **27**, 148.

Rachkovsky, D.N. (1962). Magnetic rotation effects in spectral lines (in Russian). *Izv. Krymsk. Astrofiz. Obs.* **28**, 259.

Rachkovsky, D.N. (1967). The reduction for anomalous dispersion in the theory of the absorption line formation in a magnetic field (in Russian). *Izv. Krymsk. Astrofiz. Obs.* **37**, 56.

Stenflo, J.O. (1991). Unified classical theory of line formation in a magnetic field, in *Solar polarimetry*. L.J. November (ed.) (National Solar Observatory/ Sacramento Peak: Sunspot, NM), 416.

Stenflo, J.O. (1994). *Solar magnetic fields. Polarized radiation diagnostics* (Kluwer Academic Publishers: Dordrecht). Chapter 3.

Unno, W. (1956). Line formation of a normal Zeeman triplet. *Pub. Astron. Soc. Japan*, **8**, 108.

8

The RTE in the presence of a magnetic field

Magnetic fields are to astrophysics as sex is to psychology.

—*H. C. van de Hulst, 1989.*

Now that we have formulated the general RTE for a stratified anisotropic medium in LTE, let us particularize our study to the case of an atomic vapor permeated by a magnetic field, $\boldsymbol{B}$. For convenience, we shall consider the medium to be isotropic in the absence of an "external" magnetic field. It is thus $\boldsymbol{B}$ that establishes the optical anisotropy by introducing a "preferential" direction.

In order to understand the basic concepts, we start again with the simple Lorentz model of the electron as in Chapter 6 (this time introducing the Lorentz force in the dynamical balance). In this way, the so-called "normal" Zeeman effect gets fully explained. The "anomalous" Zeeman effect, however, needs further results from quantum mechanics that will be summarized later. As the reader may already have realized from the historical introduction, this procedure conforms with historical developments. As in many other branches of physics, a chronological treatment helps in comprehension, although it is not strictly necessary.

In this chapter, we shall see that a single (unpolarized) spectral line in the absence of a magnetic field splits into various Zeeman components, each with a distinct state of polarization that may, of course, vary along the profile.

8.1 The Lorentz model of the electron

Let us resume our discussion of Section 6.2 on the Lorentz model. If the medium is assumed to be isotropic, Newton's second law for each of the three components of the electron position [Eq. (6.23)] now reads:

$$m\ddot{r}_\alpha(t) = -\mathrm{e}_0 E_\alpha(t) - q r_\alpha(t) - m\gamma\dot{r}_\alpha(t), \tag{8.1}$$

where $q = q_\alpha$ and $\gamma = \gamma_\alpha$, $\forall\,\alpha = +1, 0, -1$. The solution of this equation and the ensuing developments of Chapter 6 lead to single (i.e., isotropic) normalized absorption and dispersion profiles given by

$$\phi(u_0, a) = \frac{1}{\sqrt{\pi}} H(u_0 - u_{\mathrm{LOS}}, a) \tag{8.2}$$

and

$$\psi(u_0, a) = \frac{1}{\sqrt{\pi}} F(u_0 - u_{\mathrm{LOS}}, a) \tag{8.3}$$

because the resonant frequencies $u_{0,\alpha}$ and the damping coefficients a_α are all three equal to u_0 and a, respectively.

Assume now that a magnetic field $\boldsymbol{B} = B\hat{\boldsymbol{e}}_0$ is applied. A new force, the Lorentz force,

$$\boldsymbol{F}_{\mathrm{L}} = -\frac{\mathrm{e}_0}{c}\dot{\boldsymbol{r}} \wedge \boldsymbol{B},$$

is then exerted on the electron and must be included in the equation of motion. Here, the symbol $\wedge$ represents the vector product, and the equation is written in the Gaussian system of units. Note that the assumption that $\boldsymbol{B} \parallel \hat{\boldsymbol{e}}_0$ is completely general. If the medium is originally isotropic, all directions are equivalent. It is indeed the magnetic field that makes $\hat{\boldsymbol{e}}_0$ a "preferential" direction, hence establishing anisotropy. With this assumption, light polarized along the $\hat{\boldsymbol{e}}_0$ direction is *linearly polarized along the magnetic field lines* and light polarized in the the $\hat{\boldsymbol{e}}_{-1}$ ($\hat{\boldsymbol{e}}_{+1}$) direction is *right-hand (left-hand) circularly polarized in a plane perpendicular to the magnetic field.*

The α component of the Lorentz force is

$$F_{\mathrm{L},\alpha} = \boldsymbol{F}_{\mathrm{L}} \cdot \hat{\boldsymbol{e}}_\alpha^* = -\mathrm{i}\,\frac{\mathrm{e}_0 B}{c}\,\alpha\,\dot{r}_\alpha,$$

where we have used the vector product $\hat{\boldsymbol{e}}_\alpha \wedge \hat{\boldsymbol{e}}_0 = \mathrm{i}\alpha\hat{\boldsymbol{e}}_\alpha$, $\forall\,\alpha = +1, 0, -1$. Adding this term to Eq. (8.1) and following the same steps as in Section 6.2, we arrive at

$$\begin{aligned}\delta_\alpha &= \frac{N\mathrm{e}_0^2}{4\pi m \nu}\frac{\nu_{0,\alpha} - \nu}{(\nu_{0,\alpha} - \nu)^2 + \Gamma^2}, \\ \kappa_\alpha &= \frac{N\mathrm{e}_0^2}{4\pi m \nu}\frac{\Gamma}{(\nu_{0,\alpha} - \nu)^2 + \Gamma^2},\end{aligned} \tag{8.4}$$

so that the real and imaginary parts of the refractive index have the same general expressions [Eqs (6.29)] but now with an isotropic damping coefficient, with resonant frequencies given by

$$\nu_{0,\alpha} = \nu_0 + \alpha\nu_{\mathrm{L}}, \quad \alpha = +1, 0, -1, \tag{8.5}$$

and with resonant wavelengths given by

$$\lambda_{0,\alpha} = \lambda_0 - \alpha\lambda_B, \quad \alpha = +1, 0, -1, \tag{8.6}$$

where ν_0 and λ_0 are the (isotropic) resonant frequency and wavelength in the absence of $\boldsymbol{B}$, ν_{L} is the Larmor frequency,

$$\nu_{\mathrm{L}} \equiv \frac{e_0 B}{4\pi mc}, \tag{8.7}$$

and

$$\lambda_B \equiv \frac{\lambda_0^2 \nu_{\mathrm{L}}}{c} \tag{8.8}$$

is the Zeeman wavelength splitting†, which can be expressed numerically as

$$\lambda_B = 4.67 \times 10^{-13} \lambda_0^2 B, \tag{8.9}$$

where wavelengths are expressed in angstroms and magnetic field strengths in gauss.

Let us proceed further by accounting for the quantum mechanical correction concerning the oscillator strength (Section 6.3) and the thermal and macroscopic motions of material particles (Sections 6.4 and 6.5). If we set

$$u_0 \equiv \frac{\nu_0 - \nu}{\Delta\nu_{\mathrm{D}}} = \frac{\lambda - \lambda_0}{\Delta\lambda_{\mathrm{D}}} \tag{8.10}$$

and

$$u_B \equiv \frac{\nu_{\mathrm{L}}}{\Delta\nu_{\mathrm{D}}} = \frac{\lambda_B}{\Delta\lambda_{\mathrm{D}}}, \tag{8.11}$$

the absorption and dispersion profiles become

$$\phi_\alpha(u_0, a) = \frac{1}{\sqrt{\pi}} H(u_0 + \alpha u_B - u_{\mathrm{LOS}}, a) \tag{8.12}$$

and

$$\psi_\alpha(u_0, a) = \frac{1}{\sqrt{\pi}} F(u_0 + \alpha u_B - u_{\mathrm{LOS}}, a). \tag{8.13}$$

Note that the single (unpolarized) absorption and dispersion profiles in the absence of $\boldsymbol{B}$ are each now split into three components that are shifted in frequency with respect to the original position. The right-handed circular component (-1) is shifted to the red (longer wavelengths) and the left-handed circular component $(+1)$ to the blue (shorter wavelengths). The linearly polarized component (0) in the direction of the magnetic field is unperturbed from the original position in the

† In fact, λ_B is the splitting of a normal Zeeman triplet of Landé factor unity. It is helpful to describe general Zeeman patterns when it is used as a unit of measure (the Lorentz unit). See Section 8.3.1.

absence of the field. Note that Eq. (8.9) tells us that *the greater the magnetic field strength (or the longer the wavelength), the wider the splitting.* Therefore, for a given spectral line, the amount of splitting may serve as a diagnostic of the magnetic field strength. Also, for a given magnetic field strength permeating the atomic vapor, spectral lines at longer wavelengths provide more accurate diagnostics (e.g., the splitting in the infrared is greater than in the visible).

We can label the three components with subscripts "r" (for "red" or for "right-handed"), "b" (for "blue"), and "p" (for "principal"), respectively. With this notation, the elements of the propagation matrix read

$$\begin{aligned}
\eta_I &= 1 + \frac{\eta_0}{2}\left\{\phi_{\rm p}\sin^2\theta + \frac{1}{2}[\phi_{\rm b} + \phi_{\rm r}](1+\cos^2\theta)\right\},\\
\eta_Q &= \frac{\eta_0}{2}\left\{\phi_{\rm p} - \frac{1}{2}[\phi_{\rm b} + \phi_{\rm r}]\right\}\sin^2\theta\cos 2\varphi,\\
\eta_U &= \frac{\eta_0}{2}\left\{\phi_{\rm p} - \frac{1}{2}[\phi_{\rm b} + \phi_{\rm r}]\right\}\sin^2\theta\sin 2\varphi,\\
\eta_V &= \frac{\eta_0}{2}[\phi_{\rm r} - \phi_{\rm b}]\cos\theta,
\end{aligned} \tag{8.14}$$

and

$$\begin{aligned}
\rho_Q &= \frac{\eta_0}{2}\left\{\psi_{\rm p} - \frac{1}{2}[\psi_{\rm b} + \psi_{\rm r}]\right\}\sin^2\theta\cos 2\varphi,\\
\rho_U &= \frac{\eta_0}{2}\left\{\psi_{\rm p} - \frac{1}{2}[\psi_{\rm b} + \psi_{\rm r}]\right\}\sin^2\theta\sin 2\varphi,\\
\rho_V &= \frac{\eta_0}{2}[\psi_{\rm r} - \psi_{\rm b}]\cos\theta,
\end{aligned} \tag{8.15}$$

where the angles θ and φ have obviously to be interpreted as the inclination angle of the magnetic field vector with respect to the propagation direction and the azimuth angle of $\boldsymbol{B}$ with respect to the Stokes Q positive direction, respectively.

A graphical illustration of the wavelength shape of the various propagation matrix elements is presented in Fig. 8.1 for a normal triplet and for two values of the Zeeman splitting. The parameters are: $\eta_0 = 10$, $a = 0.05$, $\theta = \pi/4$, $\varphi = \pi/6$, and $u_B = 2.4$ (solid lines) and $u_B = 1.2$ (dashed lines). Note the strong dependence on the splitting, that Q and U profiles only differ in scale, and the significantly broad wings of ρ_V that demand radiative transfer calculations at a fair distance from the line core.

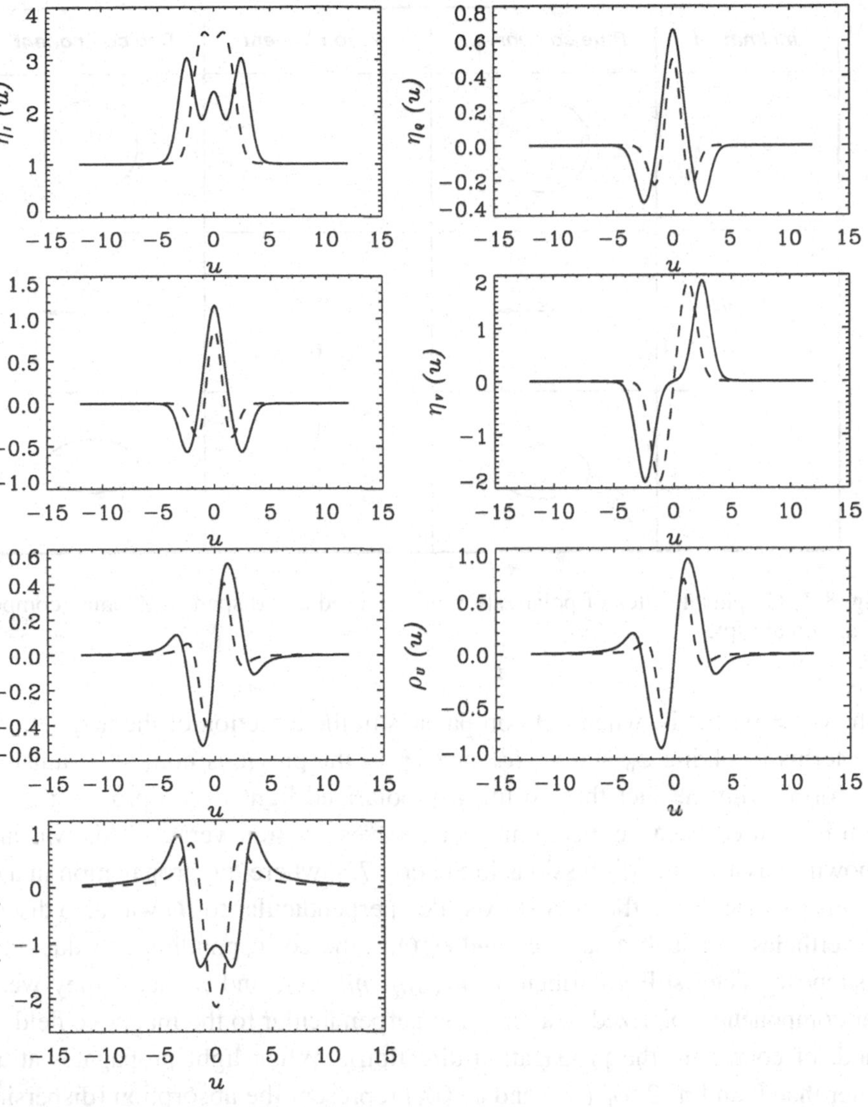

Fig. 8.1. Examples of absorption and dispersion profiles for a normal Zeeman triplet. See text for details.

The labeling of the Zeeman components according to wavelength shifts is convenient for the following reason. So far, the association between the values of index α and the polarization character of the component profiles has been stressed. Hopefully, this association has helped the reader in understanding the sign conventions. However, it might be confusing because the polarization character depends on the propagation direction. Elliptically polarized light is absorbed and dispersed in general as reflected in Eqs (8.14) and (8.15). As can easily be seen in the equations,

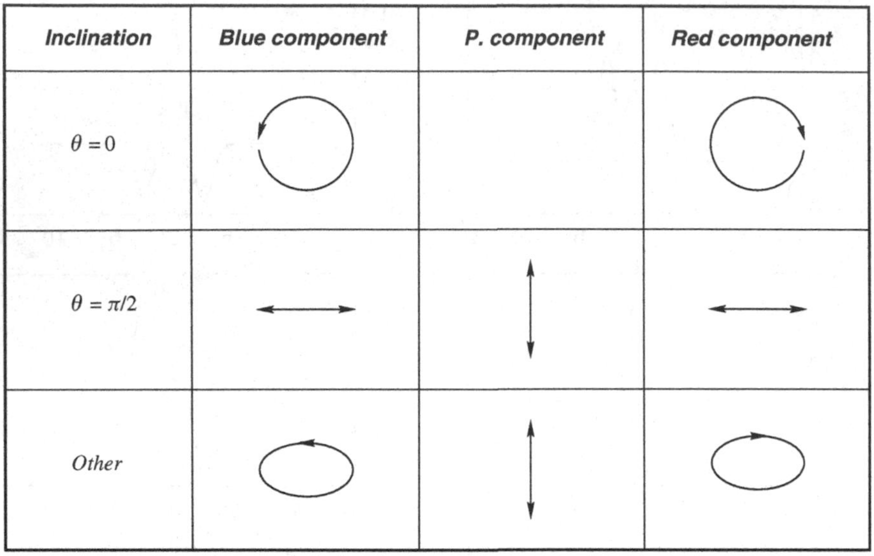

Fig. 8.2. Graphical states of polarization of absorbed and dispersed Zeeman components of a normal triplet.

when $\theta = 0$ (that is, when light propagates in the direction of the magnetic field), the terms involving ϕ_{p} and ψ_{p} (ϕ_0 and ψ_0 in the previous notation) vanish. This conforms with the fact that no linearly polarized light in the propagation direction is allowed because electromagnetic waves are transversal. This was already known to us after the discussions in Section 7.8, where the propagation in a direction perpendicular to the optical axis (i.e., perpendicular to $\boldsymbol{B}$) was also discussed. Nevertheless, in such a case, ϕ_{b} and ϕ_{r} (and the corresponding ψ values) do not disappear. They still contribute to η_I, η_Q, η_U, ρ_Q, and ρ_U as if they were linear components polarized in a direction perpendicular to the magnetic field vector (and, of course, to the propagation direction). When light propagates at angles other than 0 and $\pi/2$, ϕ_{b} (ψ_{b}) and ϕ_{r} (ψ_{r}) represent the absorption (dispersion) of elliptically polarized light, as shown in Fig. 8.2.

8.1.1 Symmetry properties of the propagation matrix elements

The elements of the propagation matrix shown in Fig. 8.1 have remarkable symmetry properties. Note that η_I, η_Q, η_U, and ρ_V are symmetric in wavelength with respect to the central position of the line;† η_V, ρ_Q, and ρ_U are antisymmetric. More explicitly,

† We understand that the shift u_{LOS} due to the macroscopic motions of the material is already taken into account. Hence, the central position of the line is given here by $u_0 - u_{\mathrm{LOS}}$.

$$\begin{aligned}
\eta_I(u) &= \eta_I(-u), \\
\eta_Q(u) &= \eta_Q(-u), \\
\eta_U(u) &= \eta_U(-u), \\
\eta_V(u) &= -\eta_V(-u), \\
\rho_Q(u) &= -\rho_Q(-u), \\
\rho_U(u) &= -\rho_U(-u), \\
\rho_V(u) &= \rho_V(-u).
\end{aligned} \tag{8.16}$$

These symmetry properties do not occur by chance, but are a direct consequence of Eqs (8.12), (8.13), (8.14), and (8.15). In fact, Eq. (8.12) implies that

$$\phi_{\rm r}(u) = \phi_{-1}(u) = \phi_{+1}(-u) = \phi_{\rm b}(-u), \tag{8.17}$$

and Eq. (8.13) that

$$\psi_{\rm r}(u) = \psi_{-1}(u) = -\psi_{+1}(-u) = -\psi_{\rm b}(-u). \tag{8.18}$$

Therefore,

$$\begin{aligned}
\phi_{\rm b}(u) + \phi_{\rm r}(u) &= \phi_{\rm r}(-u) + \phi_{\rm b}(-u), \\
\psi_{\rm r}(u) - \psi_{\rm b}(u) &= -\psi_{\rm b}(-u) + \psi_{\rm r}(-u),
\end{aligned} \tag{8.19}$$

and

$$\begin{aligned}
\phi_{\rm r}(u) - \phi_{\rm b}(u) &= \phi_{\rm b}(-u) - \phi_{\rm r}(-u), \\
\psi_{\rm b}(u) + \psi_{\rm r}(u) &= -\psi_{\rm r}(-u) - \phi_{\rm b}(-u).
\end{aligned} \tag{8.20}$$

These two last relationships, together with Eqs (8.14) and (8.15), readily give the expressions (8.16).

8.2 *LS* coupling

The simple Lorentz model of the electron is able to explain only the shape of Zeeman triplet profiles. Experience, however, reveals a large number of Zeeman multiplets for which classical physics can provide no explanation. A quantum-mechanical treatment is therefore mandatory. We summarize such a treatment here, although details of the rigorous theory are beyond the scope of this book. Let us start to describe the model atom by finding those quantum observables that provide the right framework for evaluating the energy jumps associated with the atomic transitions that produce spectral lines.

As in classical mechanics, the total angular momentum of an n-particle system with respect to a fixed point in space is a constant of the motion in the absence of external forces. This is not, of course, the case for individual angular momenta because of internal forces (interactions) among the particles. Such internal interactions induce transfer of angular momentum from one particle to another. It is therefore important to know the total angular momentum of the system in order to find the system eigenstates. The addition of individual angular momenta, however, cannot be made in just one way. Consider an atom with n electrons, each with an orbital angular momentum $\boldsymbol{l}_i$ and with a spin angular momentum $\boldsymbol{s}_i$. The angular momenta are assumed to be calculated with respect to the central position of the (infinitely) massive nucleus, which is at rest. The description of the system is different if one first adds $\boldsymbol{l}_i$ and $\boldsymbol{s}_i$ to get $\boldsymbol{j}_i$ and then all the $\boldsymbol{j}_i$'s than if one proceeds by adding first all the $\boldsymbol{l}_i$'s on the one hand, all the $\boldsymbol{s}_i$'s on the other, and then calculating the sum of both. An ordering must then be chosen that reflects as accurately as possible the various degrees of importance or strengths of the different internal interactions.

In our atomic system we can distinguish three main interactions. The Coulombic repulsion between every two electrons certainly couples their individual orbital angular momenta. The spin–spin interaction among individual electrons, owing to spin statistics, couples individual spin angular momenta. The spin–orbit interaction, owing to the magnetic field "seen" by the electron that moves in the electrostatic electric field of the nucleus (a special-relativity effect), couples the orbital with the individual spin angular momenta. Assuming the relative strengths of these three interactions to be in that order, that is, assuming that the Coulomb interaction is the most important, it seems reasonable to proceed by first evaluating a total orbital angular momentum,

$$\boldsymbol{L} \equiv \sum_{i=1}^{n} \boldsymbol{l}_i, \tag{8.21}$$

then a total spin angular momentum,

$$\boldsymbol{S} \equiv \sum_{i=1}^{n} \boldsymbol{s}_i, \tag{8.22}$$

and finally the total angular momentum,

$$\boldsymbol{J} \equiv \boldsymbol{L} + \boldsymbol{S}. \tag{8.23}$$

This coupling scheme is known as the *LS* or Russell–Saunders coupling and is the one we shall use hereafter. It is a simple coupling scheme because it permits the same magnetic perturbation to the system Hamiltonian as that of a one-electron system. It is also the most extensively used in astrophysics. In several cases (in particular for neutral iron lines of high excitation potential) *LS* coupling fails and it becomes necessary to resort to another coupling scheme. Nevertheless, the differences among the results of the various coupling schemes are more of a quantitative than a qualitative character and the basic physics remains almost the same.

If the atomic conditions are close to those of *LS* coupling, the observables $\mathbf{L}^2$, $\mathbf{S}^2$, $\mathbf{J}^2$, and J_z turn out to form a complete set of commuting operators, so that their associated quantum numbers l, s, j, and m, respectively, are *good* quantum numbers and states $|lsjm\rangle$ characterized by them are eigenvectors of all four observables:

$$\begin{aligned} \mathbf{L}^2|lsjm\rangle &= \hbar^2\, l(l+1)\, |lsjm\rangle, \\ \mathbf{S}^2|lsjm\rangle &= \hbar^2\, s(s+1)\, |lsjm\rangle, \\ \mathbf{J}^2|lsjm\rangle &= \hbar^2\, j(j+1)\, |lsjm\rangle, \\ \mathrm{J}_z|lsjm\rangle &= \hbar\, m\, |lsjm\rangle \end{aligned} \tag{8.24}$$

where $\hbar$ is Planck's constant, h, divided by 2π.

All these quantum states are also energy eigenstates because, in the absence of an external field, the Hamiltonian, $\mathbf{H}_0$, also commutes with the total angular momentum as a consequence of its rotational invariance. Remarkably, all quantum states with quantum number, j, have the same eigenvalue, E_j,

$$\mathbf{H}_0\, |lsjm\rangle = E_j\, |lsjm\rangle, \tag{8.25}$$

regardless of m. We say that the state is $(2j+1)$-fold degenerate, or that the energy level has $2j+1$ sub-levels each characterized by the m quantum number $(m = -j, -j+1, \ldots, 0, \ldots, j-1, j)$.

8.3 The Zeeman effect

If an external magnetic field is applied to the system, a new Hamiltonian term must be added to $\mathbf{H}_0$ to account for the interaction energy between the atom and $\boldsymbol{B}$. Assume that this new term is just a small perturbation to the energy levels of the atom in the absence of the magnetic field. That is, the matrix elements of the new term, $\mathbf{H}_B$, are supposed to be small when compared with those of $\mathbf{H}_0$. It is then necessary just to evaluate those $\mathbf{H}_B$ matrix elements and add them to the eigenvalues of the unperturbed system.

If the magnetic field is homogeneous within the spatial domain of atomic dimensions, the magnetic Hamiltonian is given by

$$\mathbf{H}_B = \boldsymbol{\mu} \cdot \boldsymbol{B} + O(B^2), \tag{8.26}$$

where $\boldsymbol{\mu}$ is the atom's intrinsic magnetic moment†

$$\boldsymbol{\mu} = \mu_0(\boldsymbol{J} + \boldsymbol{S}) \tag{8.27}$$

and the term $O(B^2)$ is the so-called diamagnetic term (of the order of B^2, but not

† We assume that the spin gyromagnetic ratio is $g_s = 2$. In fact, experimental measurements and quantum electrodynamics give $g_s = 2\,[1 + \alpha/\pi + O(\alpha/\pi)^2] \simeq 2.0023192$, where α is the fine structure constant.

explicitly written), which is perfectly negligible for our purposes when dealing with small enough magnetic field strengths like those usually produced in the laboratory or found in the Sun and most stars.†

The quantity μ_0 is the Bohr magneton,

$$\mu_0 \equiv \frac{e_0 \hbar}{2mc} = \frac{h\nu_L}{B} = 9.27 \times 10^{-21} \text{ erg G}^{-1}. \quad (8.28)$$

The diagonal terms of $\mathbf{H}_B$ turn out to be

$$\langle lsjm|\mathbf{H}_B|lsjm\rangle = mg\mu_0 B = mg\, h\nu_L, \quad (8.29)$$

where g is the *Landé factor* of the level. In *LS* coupling, the Landé factor is given by

$$g_{LS} = \frac{3}{2} + \frac{s(s+1) - l(l+1)}{2j(j+1)} \quad (8.30)$$

when $j \neq 0$. Note that when $j = 0$ the Landé factor loses its meaning since $m = 0$ and the magnetic perturbation is zero.

The non-diagonal terms are not zero, however. $\mathbf{H}_B$ is diagonal for all quantum numbers except for j. The total angular momentum is no longer an invariant of the motion, although J_z is still invariant.

$$\langle lsjm|\mathbf{H}_B|l's'j'm'\rangle = -\delta_{lsmj-1,l's'm'j'}$$
$$\times \left[\frac{(j^2 - m^2)(j + l + s + 1)(j + l - s)(j + s - l)(l + s - j + 1)}{4j^2(2j - 1)(2j + 1)}\right] h\nu_L. \quad (8.31)$$

Fortunately, these non-diagonal terms are negligible with respect to the diagonal ones for the range of weak magnetic fields we are dealing with, so that we can still consider $\boldsymbol{J}$ as an approximate motion invariant and the states $|lsjm\rangle$ as eigenvectors of the Hamiltonian:

$$(\mathbf{H}_0 + \mathbf{H}_B)\,|lsjm\rangle = (E_j + mg\, h\nu_L)\,|lsjm\rangle. \quad (8.32)$$

Therefore, the degeneracy of the energy level j has been broken up by the presence of the magnetic field into $2j + 1$ components whose shifts in frequency (energy) are proportional to the (magnetic) quantum number m and to the Larmor frequency, ν_L (see Fig. 8.3). The adjective "magnetic" applied to quantum number m comes from the customary selection of the third component of the total angular momentum along the magnetic field direction. Hence, the J_z eigenvalue turns out

† Very strong magnetic fields have been reported on white dwarfs and pulsars, for which the diamagnetic term might be needed. The order of magnitude of $O(B^2)$ with respect to $\boldsymbol{\mu}\cdot\boldsymbol{B}$ is the same as that of the latter (paramagnetic term) with respect to the unperturbed Hamiltonian, $\mathbf{H}_0$.

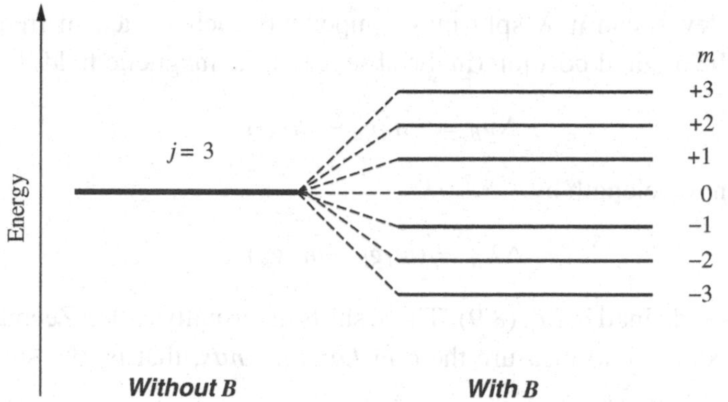

Fig. 8.3. A level with quantum number $j = 3$ is split in the presence of a magnetic field into seven sub-levels, each shifted proportionally to their magnetic quantum number, m.

to be the projection of the total angular momentum in the magnetic field direction. Note that the proportionality coefficient, the Landé factor, has been written without the *LS* subscript. This makes the result more general since, when *LS* coupling is not a suitable scheme, Landé factors can be calculated within other coupling schemes or even be determined empirically from laboratory experiments (see later).

8.3.1 Allowed atomic transitions

We understand the formation of spectral lines as a consequence of atomic transitions between two levels l (for lower) and u (for upper) whose energies are $E_l < E_u$ with an ensuing absorption or emission of a photon. With this convention, a transition from the upper to the lower level is an emission line whereas a transition from the lower to the upper level is an absorption line. Either in the absence or in the presence of a magnetic field, conservation of angular momentum demands that $\boldsymbol{J}_u = \boldsymbol{J}_l + \boldsymbol{J}_\gamma$. Since the photon has $j_\gamma = 1$, according to the quantum-mechanical rule for addition of angular momenta,

$$\Delta j \equiv j_u - j_l = 0, \ \pm 1 \tag{8.33}$$

except for the case

$$j_u = j_l = 0,$$

which is forbidden.

According to Eq. (8.32), when a magnetic field is present the level energies depend not only on j but on m as well. So what was a single line between two degenerate levels can now split into components each shifted in frequency with respect to the original position (in the absence of the magnetic field) by

$$\Delta \nu_B = (m_u g_u - m_l g_l)\, \nu_{\mathrm{L}}, \tag{8.34}$$

or shifted in wavelength by

$$\Delta \lambda_B = (m_l g_l - m_u g_u)\, \lambda_B, \tag{8.35}$$

where λ_B was defined in Eq. (8.9). These shifts are usually called Zeeman splittings and it is customary to measure them in *Lorentz units*, that is, the shift is simply evaluated as $\Delta\lambda_B/\lambda_B$.

Now the questions arise as to how many of the possible transitions represent physically allowed transitions and what their polarization states are. The answers are in fact very easy and again come from the conservation of angular momentum.

The only three possible values for the photon helicity (the equivalent to the magnetic quantum number m_γ) in the atomic reference system are 0 and ± 1. These three values correspond to linearly polarized photons in the $\boldsymbol{B}$ direction ($m_\gamma = 0$), and to right-handed ($m_\gamma = -1$) and left-handed ($m_\gamma = +1$) circularly polarized radiation in a plane perpendicular to $\boldsymbol{B}$. Therefore, there are just three possibilities for allowed jumps of the magnetic quantum number between the two levels of a transition, namely,

$$\Delta m \equiv m_u - m_l = 0,\ \pm 1. \tag{8.36}$$

Equations (8.33) and (8.36) are known as the selection rules for electric dipole transitions and, as a matter of fact, can be rigorously derived within the quantum-mechanical formalism. Other selection rules apply to different transition types (magnetic dipole, electric quadrupole, etc.) and new rules may be added to electric dipole transitions in strict *LS* coupling conditions. Nonetheless, these two rules, Eqs (8.33) and (8.36), provide results that are valid even when the atomic conditions are far from those for *LS* coupling. We will therefore not discuss selection rules further.

8.3.2 The Zeeman pattern

It can readily be understood that these three values of Δm correspond to the three values of index α of Section 8.1 and Chapters 6 and 7. Then we can make the correspondence between the so-called π transitions ($\Delta m = 0$) and the normalized absorption and dispersion profiles ϕ_{p} and ψ_{p}; between the so-called σ_{b} transitions ($\Delta m = +1$) and the profiles ϕ_{b} and ψ_{b}; and between the so-called σ_{r} transitions ($\Delta m = -1$) and the profiles ϕ_{r} and ψ_{r}. The considerations of propagation of elliptically polarized components summarized in Fig. 8.2 are still valid. Nevertheless,

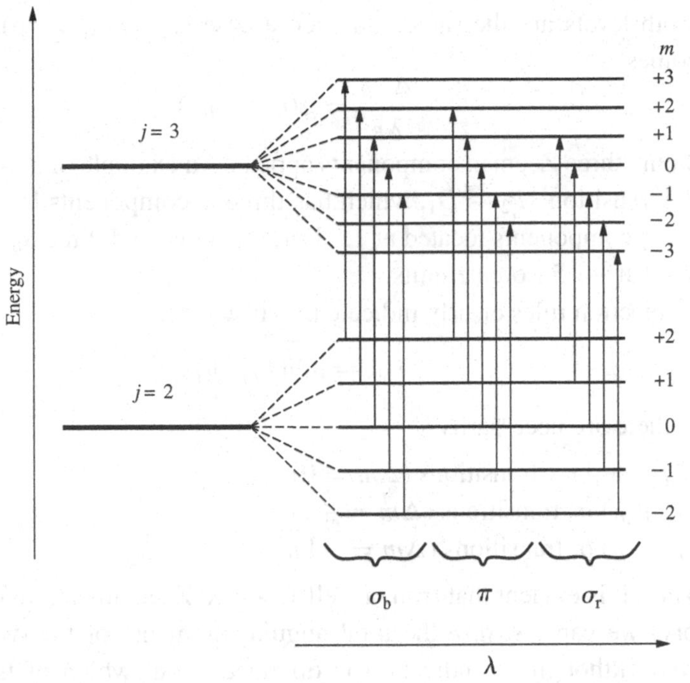

Fig. 8.4. What was a single line in the absence of a magnetic field is split into a pattern of $(2j_{\min} + 1)$ π components, $(j_u + j_l)$ σ_b components, and $(j_u + j_l)$ σ_r components.

a number of differences between this general development and that of a normal Zeeman triplet are worth noting:

(1) The number of components is not always three. A pattern of several components appears (see Fig. 8.4). Classical theory (i.e., Lorentz's model) was only able to explain a triplet whose Zeeman splitting is unity (in Lorentz units). Such a triplet was called a "normal" Zeeman pattern and any other was labeled "anomalous", hence the frequent references to the normal and anomalous Zeeman effects. Our selection rules indicate, however, that the adjective normal refers mostly to the exception rather than to the rule. As a matter of fact, Eqs (8.33), (8.36), and (8.34) or (8.35) explain that a normal Zeeman pattern can be produced between two levels of total angular momentum 0 and 1, the latter having a Landé factor of unity. Otherwise, the pattern is said to be anomalous. If the total angular momenta are 0 and 1 but the latter has a Landé factor different from unity, a triplet is produced as well but with a different splitting. Another interesting case for which a triplet is obtained is that of the so-called pseudo triplet: the two total angular momenta can have any allowed values but the Landé factors

of both levels are the same. In such a case ($g_u = g_l = g$), Eq. (8.35) becomes

$$\frac{\Delta\lambda_B}{\lambda_B} = g(m_l - m_u)$$

and only three Zeeman components appear. An example of a pseudo-triplet is the transition ${}^3P_2$—${}^5D_1$, which has three π components located at zero, three σ_r components located at 1.5 Lorentz units, and three σ_b components located at -1.5 Lorentz units.

(2) The selection rules clearly indicate that if we call

$$j_{min} \equiv \min\{j_u, j_l\}, \tag{8.37}$$

then there are necessarily

- $(2j_{min} + 1)$ π transitions ($\Delta m = 0$),
- $(j_u + j_l)$ σ_b transitions ($\Delta m = +1$),
- $(j_u + j_l)$ σ_r transitions ($\Delta m = -1$).

Thus, it is evident that from a well-resolved Zeeman pattern (in the laboratory) we can *measure* the total angular momenta of the two levels involved, although no indication is obtained about which of them corresponds to the upper level and which to the lower level. This is not much of a problem because remarkable symmetry properties exist that imply the same Zeeman pattern if one interchanges upper and lower levels. An example Grotrian diagram of a Zeeman pattern between two levels of $j_l = 2$ and $j_u = 3$ can be seen in Fig. 8.4.

(3) Both the π and the σ patterns are symmetric about the central original position of the line in the absence of magnetic fields.

The wavelength Zeeman splitting in Lorentz units is,

(a) for π transitions ($m_l = m_u \equiv m$),

$$\frac{\Delta\lambda_B}{\lambda_B} = m\,(g_l - g_u), \tag{8.38}$$

(b) for σ_b transitions ($m_u = m_l + 1$),

$$\frac{\Delta\lambda_B}{\lambda_B} = m_l\,(g_l - g_u) - g_u = m_u\,(g_l - g_u) - g_l, \tag{8.39}$$

(c) and for σ_r transitions ($m_u = m_l - 1$),

$$\frac{\Delta\lambda_B}{\lambda_B} = m_l\,(g_l - g_u) + g_u = m_u\,(g_l - g_u) + g_l. \tag{8.40}$$

For every π transition with $m \neq 0$ there is another with $m' = -m$ whose splitting is the opposite. Therefore, the π spectrum is symmetric about zero.

For every σ_b transition with magnetic quantum number m_l there is a σ_r transition with $m_l' = -m_l$ and vice versa. Hence, Eqs (8.39) and (8.40) imply that every σ component has a symmetric counterpart with respect to the central position.

(4) From Eqs (8.38), (8.39), and (8.40), it is evident that the separation, d, between two adjacent Zeeman components of both the π spectrum and either half (σ_b or σ_r) of the σ spectrum is

$$d = |g_u - g_l|. \tag{8.41}$$

Let us consider separately the two cases $\Delta j = 0$ and $|\Delta j| = 1$. When $j_u = j_l = j$, the magnetic quantum number of the upper level runs from $-j + 1$ to j for the σ_b transitions and from $-j$ to $j - 1$ for the σ_r transitions. According to Eqs (8.39) and (8.40), the mid-points of the two halves of the σ spectrum are located at $-(g_u + g_l)/2$ and at $(g_u + g_l)/2$. Hence, the distance, d_σ, between such mid-points is

$$d_\sigma = g_u + g_l. \tag{8.42}$$

When $|\Delta j| = 1$, it is convenient to write Eqs (8.39) and (8.40) in terms of m_min, g_min, m_max, and g_max, where the subscripts now refer to the level with smaller or greater j quantum number. It is easy to see that, in this case, m_min runs from $-j_\text{min}$ to j_min for both the σ_b and the σ_r transitions, so that the distance between the mid-points of the σ patterns is

$$d_\sigma = 2g_\text{max}. \tag{8.43}$$

All three results taken together allow us to determine the two Landé factors empirically. This turns out to be of great help especially for lines that are formed in conditions far from *LS* coupling. In any case, empirically measured Landé factors may help in discovering errors or discrepancies between lines that have previously been assumed in *LS* coupling (Landi Degl'Innocenti, 1982; Stenflo *et al.*, 1984; Solanki and Stenflo, 1985).

(5) The Landé factor for *LS* coupling (8.30) may be negative. This circumstance heralds the possibility that, eventually, some σ_b components may be shifted to the red and the symmetric σ_r components to the blue. This is a completely new effect that was not foreseen in the Lorentz model. Moreover, this is not a particular feature of *LS* coupling but a general result that can be understood better from Eqs (8.39) and (8.40), which are valid for all atomic coupling conditions. Consider, for example, Eq. (8.39) as if it represents a straight line: the Zeeman splitting takes different (discrete) values along a straight line of slope $g_l - g_u$ and offset $-g_u$ depending on the values of m_l.

Such a straight line crosses $\Delta\lambda_B/\lambda_B = 0$ at the point

$$m^* = \frac{g}{g_l - g_u}.$$

If $g_u > 0$ and m^* is outside the allowed interval for m_l, then all σ_b Zeeman splittings will be negative (and all σ_r Zeeman splittings will be positive) as one would expect. Otherwise, some components may display the opposite splitting. The symmetry of the Zeeman pattern is not broken, however. An example of such a Zeeman pattern is that of the Fe II (ionized iron) line at 436.94 nm. Using the standard spectroscopic notation, the line is a ${}^4P_{1/2}$—${}^4F_{3/2}$ transition, so that $s_l = s_u = 1/2, l_l = 1, l_u = 3, j_l = 1/2$, and $j_u = 3/2$. In *LS* coupling, the Landé factors of the lower and the upper level are $g_l = 8/3$ and $g_u = 2/5$. With these Landé factors, the Zeeman splittings of the two σ_b components are (in Lorentz units) −23/15 and 11/15: one is shifted to the blue and the other to the red. This line seems not to be *LS* coupled, but the actual qualitative behavior is the same (Solanki and Stenflo, 1985).

8.3.3 Relative intensities of the Zeeman components

All we need to know for evaluating the general absorption and dispersion profiles corresponding to a general (anomalous) Zeeman pattern is the relative strengths of the various Zeeman components. These intensities are proportional to the transition probabilities between the sub-levels due to the action of the electric dipole operator. If **D** is such an operator, the probabilities are proportional to the population of the initial sub-level (say, the lower sub-level for absorption lines) and to the square modulus of its matrix elements, $|\langle lsjm|\mathbf{D}|l's'j'm'\rangle|^2$. In normal excitation conditions, i.e., in thermodynamic equilibrium, all the sub-levels are equally populated and the probabilities depend only on the quantum numbers involved. (Population imbalances as well as coherences between the different sub-levels appear, for instance, when matter is illuminated by an anisotropic radiation field, as in pure scattering processes.) Under our hypothesis of local thermodynamic equilibrium we shall be assuming that no such population imbalances and coherences apply to our atomic system. The strengths are then proportional to the above matrix elements and are given by the quantities

$$S_{\alpha,i} \equiv \frac{\tilde{S}_{\alpha,i}}{\sum_i \tilde{S}_{\alpha,i}}, \quad \alpha = +1, 0, -1, \quad \text{or b, p, r}, \tag{8.44}$$

where the unnormalized strengths, $\tilde{S}_{\alpha,i}$, are given in Table 8.1. Note that a distinction is made between transitions for which $\Delta j = 0$ and those for which $\Delta j = \pm 1$. The former have the total angular momentum of the upper level and its third component as variables. The latter transitions have the minimum of the two total

Table 8.1. *Unnormalized strengths of Zeeman components*

$j_{\min}$	$\tilde{S}_{b,i}$	$\tilde{S}_{p,i}$	$\tilde{S}_{r,i}$
$j_u = j_l$	$\frac{1}{2}(j_u+m_i)(j_u-m_i+1)$	m_i^2	$\frac{1}{2}(j_u-m_i)(j_u+m_i+1)$
j_l	$\frac{1}{2}(j_{\min}+m_i+1)(j_{\min}+m_i+2)$	$(j_{\min}+1)^2-m_i^2$	$\frac{1}{2}(j_{\min}-m_i+1)(j_{\min}-m_i+2)$
j_u	$\frac{1}{2}(j_{\min}-m_i+1)(j_{\min}-m_i+2)$	$(j_{\min}+1)^2-m_i^2$	$\frac{1}{2}(j_{\min}+m_i+1)(j_{\min}+m_i+2)$

angular momenta and its third component as variables. The expressions corresponding to blue and red components are interchanged when either the lower or the upper total angular momentum is the minimum. The denominator of Eq. (8.44) is a sum that must be extended to all Zeeman components having a jump in magnetic quantum number $\Delta m = m_u - m_l = \alpha$. According to the above description of each Zeeman pattern, the normalization conditions are:

(a)

$$\sum_{m_i=-j_u+1}^{j_u} S_{b,i} = \sum_{m_i=-j_u}^{j_u} S_{p,i} = \sum_{m_i=-j_u}^{j_u-1} S_{r,i} = 1, \tag{8.45}$$

for $\Delta j = 0$ transitions, and

(b)

$$\sum_{m_i=-j_{\min}}^{j_{\min}} S_{b,i} = \sum_{m_i=-j_{\min}}^{j_{\min}} S_{p,i} = \sum_{m_i=-j_{\min}}^{j_{\min}} S_{r,i} = 1, \tag{8.46}$$

for $\Delta j = \pm 1$ transitions.

Some examples of different Zeeman patterns are displayed in Fig. 8.5. The graphical convention used is that π components point upwards whereas σ components point downwards. From left to right and from top to bottom, the various Zeeman patterns correspond to the following transitions:† ${}^5D_0$—${}^7D_1$, ${}^5D_2$—${}^7D_3$, ${}^3P_2$—${}^5P_1$, ${}^4F_{5/2}$—${}^4F_{3/2}$, ${}^6S_{5/2}$—${}^4D_{5/2}$, and ${}^4P_{1/2}$—${}^4F_{3/2}$. Note that the first is a pure (anomalous) triplet and that the last is our example of Section 8.3.2 where one of the σ_b components (that closest to zero) is shifted to the red and one of the σ_r components (that closest to zero) is shifted to the blue.

8.4 The elements of the propagation matrix

The results of Section 8.3 provide a complete picture of what happens with matter and radiation when a spectral line is formed in the presence of a magnetic

† The atomic levels are designated with the standard spectroscopic notation ${}^{2s+1}l_{2j+1}$. The values of the orbital quantum number l are S for 0, P for 1, D for 2, F for 3, etc.

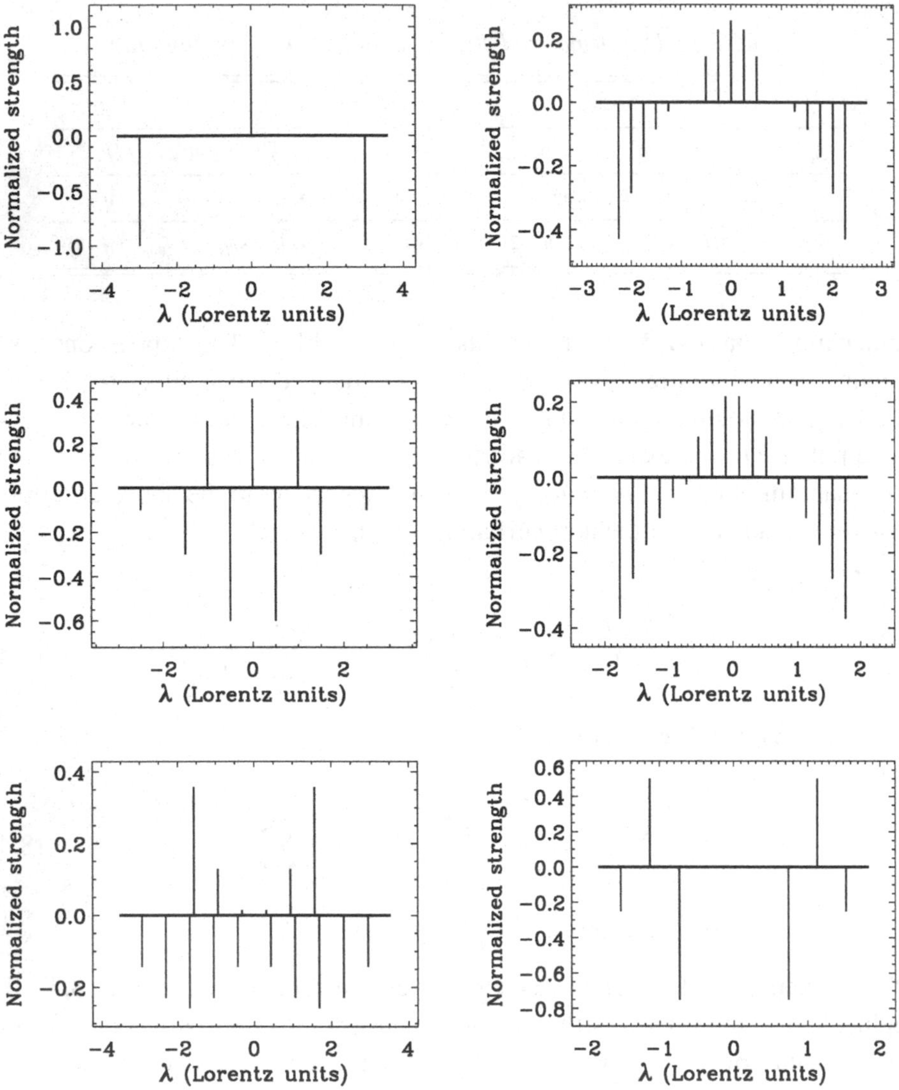

Fig. 8.5. Some examples of Zeeman patterns.

field. Each individual Zeeman component contributes to the total absorption and dispersion effects so that the elements of the propagation matrix are made up of sums of such contributions. According to Eqs (8.38), (8.39), and (8.40), the Zeeman splittings of the individual components can be written in compact form as

$$\Delta\lambda_{B,\alpha,i} = [m_{l,i}(g_l - g_u) - \alpha g_u]\,\lambda_B \tag{8.47}$$

or

$$\Delta\nu_{B,\alpha,i} = [\alpha g_u + m_{l,i}(g_u - g_l)]\,\nu_{\rm L}, \tag{8.48}$$

so that the reduced individual splittings are

$$u_{B,\alpha,i} = \frac{\Delta\nu_{B,\alpha,i}}{\Delta\nu_{\rm D}} = -\frac{\Delta\lambda_{B,\alpha,i}}{\Delta\lambda_{\rm D}}. \tag{8.49}$$

Therefore, the matrix elements of **K** remain formally the same as in Eqs (8.14) and (8.15), although the normalized profiles ϕ_α and ψ_α are now given by

$$\phi_\alpha = \frac{1}{\sqrt{\pi}} \sum_i S_{\alpha,i}\, H(u_0 + u_{B,\alpha,i} - u_{\rm LOS}, a) \tag{8.50}$$

and

$$\psi_\alpha = \frac{1}{\sqrt{\pi}} \sum_i S_{\alpha,i}\, F(u_0 + u_{B,\alpha,i} - u_{\rm LOS}, a). \tag{8.51}$$

Again, the sums must be extended to all Zeeman components having $\Delta m = m_u - m_l = \alpha$. The limits coincide with those of Eqs (8.45) and (8.46).

Figures 8.6 and 8.7 show two examples of the **K** matrix elements corresponding to ${}^3P_2$—${}^5P_1$ (left, middle panel of Fig. 8.5) and ${}^4P_{1/2}$—${}^4F_{3/2}$ (right, bottom panel of Fig. 8.5) transitions. For comparison with Fig. 8.1, the remaining parameters are $\eta_0 = 10$, $a = 0.05$, $\theta = \pi/4$, and $\varphi = \pi/6$; $\lambda_B/\Delta\lambda_{\rm D} = 2.4$ (solid lines) and 1.2 (dashed lines). As for the normal triplet case, the larger B ($\Delta\lambda_B$) or the central wavelength, the more conspicuous the contributions of the individual components. It is particularly interesting that all the matrix elements but η_I, which is always positive, qualitatively change the global sign when they are calculated for one transition or the other. This behavior follows from the essential difference between the two Zeeman patterns: remember that ${}^4P_{1/2}$—${}^4F_{3/2}$ transition is our example of a pattern having one $\sigma_{\rm b}$ component (the strongest) shifted to the red and one $\sigma_{\rm r}$ component (the strongest) shifted to the blue.

Note that the symmetry properties of every Zeeman pattern ensure that the symmetry relationships (8.16), obtained for the elements of the propagation matrix in the case of a triplet, hold generally.

8.4.1 The effective Zeeman triplet

For some applications, it may be convenient to use an approximation consisting in the substitution of the actual Zeeman pattern of a given spectral line by an "effective" Zeeman triplet. As we shall see, the usefulness of this approximation depends on the strength of the magnetic field and on the particular Zeeman pattern. However, the approximation stresses some of the concepts learnt about the general

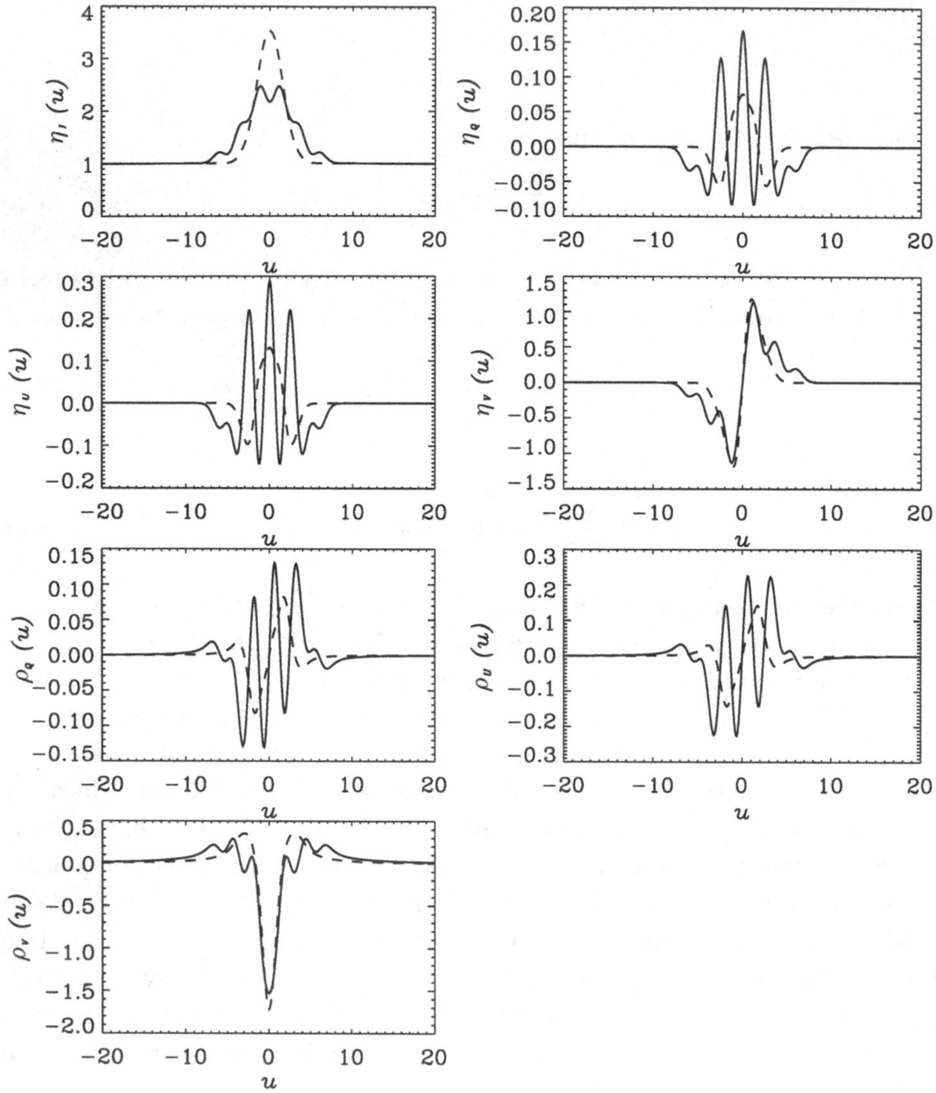

Fig. 8.6. Propagation matrix elements of the $^3P_2$—$^5P_1$ transition (see text for parameters).

Zeeman pattern and, most importantly, leads to the definition of the *effective Landé factor* of the line, a parameter of practical relevance.

Let us define

$$g_{\alpha,i} \equiv \alpha g_u + m_{l,i}(g_u - g_l). \tag{8.52}$$

The reduced splitting of each individual component [Eq. (8.49)] can then be written as

$$u_{B,\alpha,i} = g_{\alpha,i}\, u_B, \tag{8.53}$$

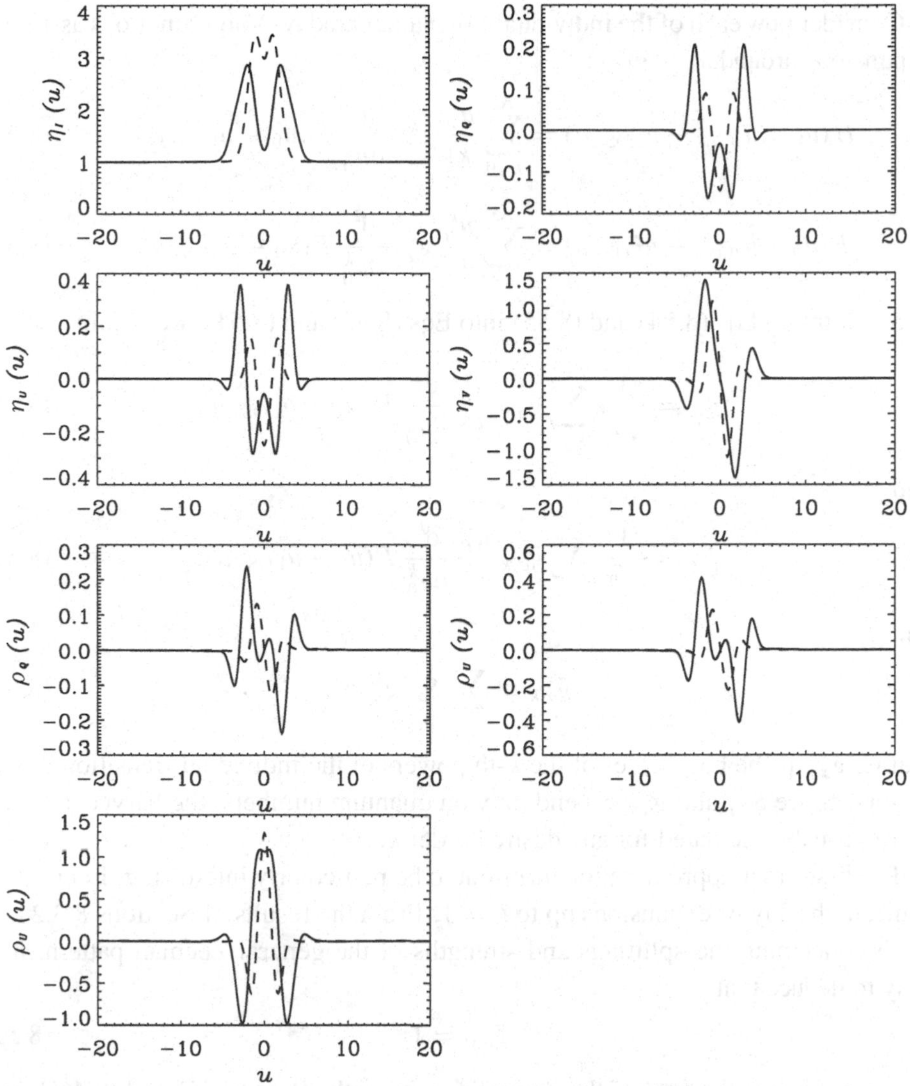

Fig. 8.7. Propagation matrix elements of the ${}^4P_{1/2}$—${}^4F_{3/2}$ transition (see text for parameters).

where u_B coincides with the splitting for a normal Zeeman triplet [Eq. (8.11)], that is,

$$u_B = \frac{\nu_L}{\Delta\nu_D} = \frac{\lambda_B}{\Delta\lambda_D}.$$

We call $g_{\alpha,i}$ the Landé factor of the individual transition.

Consider now each of the individual Voigt and Faraday–Voigt functions as Taylor expansions around $u_0 - u_{\mathrm{LOS}}$:

$$H(u_0 + u_{B,\alpha,i} - u_{\mathrm{LOS}}, a) = \sum_{k=0}^{\infty} \frac{u_B^k}{k!} g_{\alpha,i}^k \frac{\mathrm{d}^k}{\mathrm{d}u_0^k} H(u_0 - u_{\mathrm{LOS}}, a), \tag{8.54}$$

$$F(u_0 + u_{B,\alpha,i} - u_{\mathrm{LOS}}, a) = \sum_{k=0}^{\infty} \frac{u_B^k}{k!} g_{\alpha,i}^k \frac{\mathrm{d}^k}{\mathrm{d}u_0^k} F(u_0 - u_{\mathrm{LOS}}, a). \tag{8.55}$$

Substituting Eqs (8.54) and (8.55) into Eqs (8.50) and (8.51), we obtain

$$\phi_\alpha = \frac{1}{\sqrt{\pi}} \sum_{k=0}^{\infty} \bar{g}_{\alpha,k} \frac{u_B^k}{k!} \frac{\mathrm{d}^k}{\mathrm{d}u_0^k} H(u_0 - u_{\mathrm{LOS}}, a) \tag{8.56}$$

and

$$\psi_\alpha = \frac{1}{\sqrt{\pi}} \sum_{k=0}^{\infty} \bar{g}_{\alpha,k} \frac{u_B^k}{k!} \frac{\mathrm{d}^k}{\mathrm{d}u_0^k} F(u_0 - u_{\mathrm{LOS}}, a), \tag{8.57}$$

where

$$\bar{g}_{\alpha,k} \equiv \sum_i S_{\alpha,i} g_{\alpha,i}^k; \tag{8.58}$$

that is, $\bar{g}_{\alpha,k}$ is the barycenter of the k-th powers of the individual-transition Landé factors. Since $S_{\alpha,i}$ and $g_{\alpha,i}$ depend only on quantum numbers, the barycenters can be rigorously calculated for any desired order k.

The first-order approximation turns out to be particularly interesting. Let us then truncate the Taylor expansions up to $k = 1$. From the results of Sections 8.3.2 and 8.3.3 concerning the splittings and strengths of the general Zeeman pattern, it is easy to deduce that

$$\bar{g}_{\alpha,0} = 1 \tag{8.59}$$

because it is just the sum of the normalized strengths [Eqs (8.45) and (8.46)], and that

$$\bar{g}_{\mathrm{p},1} = \sum_{m_i=-j_{\min}}^{j_{\min}} S_{\mathrm{p},i}\, g_{\mathrm{p},i} = 0 \tag{8.60}$$

because the π spectrum is symmetric about zero. The quantum-mechanical calculation of $\bar{g}_{\pm 1,1}$ yields

$$\bar{g}_{\mathrm{b},1} = -\bar{g}_{\mathrm{r},1} \equiv g_{\mathrm{eff}} = \frac{1}{2}(g_u - g_l) + \frac{1}{4}(g_u - g_l)[j_u(j_u+1) - j_l(j_l+1)]. \tag{8.61}$$

The new parameter g_{eff} is called the *effective Landé factor* of the spectral line. Note that its expression is invariant under interchange of the upper and lower levels, and that it is independent of coupling schemes because so are g_u and g_l. If the latter are calculated within a given coupling scheme (e.g., *LS* coupling) the effective Landé factor can be evaluated theoretically and confronted with observations. As a matter of fact, the Sun has been used as an atomic physics laboratory for checking the validity of *LS* coupling of a number of spectral lines (Landi Degl'Innocenti, 1982; Stenflo *et al.*, 1984; Solanki and Stenflo, 1985).

Thus, to a first-order approximation, the normalized absorption and dispersion profiles look like

$$\begin{aligned}
\phi_{\mathrm{p}} &\simeq \frac{1}{\sqrt{\pi}} H(u_0 - u_{\mathrm{LOS}}, a), \\
\phi_{\mathrm{b}} &\simeq \frac{1}{\sqrt{\pi}} \left[H(u_0 - u_{\mathrm{LOS}}, a) + g_{\mathrm{eff}}\, u_B \frac{\mathrm{d}}{\mathrm{d}u_0} H(u_0 - u_{\mathrm{LOS}}, a)\right], \\
\phi_{\mathrm{r}} &\simeq \frac{1}{\sqrt{\pi}} \left[H(u_0 - u_{\mathrm{LOS}}, a) - g_{\mathrm{eff}}\, u_B \frac{\mathrm{d}}{\mathrm{d}u_0} H(u_0 - u_{\mathrm{LOS}}, a)\right],
\end{aligned} \tag{8.62}$$

and

$$\begin{aligned}
\psi_{\mathrm{p}} &\simeq \frac{1}{\sqrt{\pi}} F(u_0 - u_{\mathrm{LOS}}, a), \\
\psi_{\mathrm{b}} &\simeq \frac{1}{\sqrt{\pi}} \left[F(u_0 - u_{\mathrm{LOS}}, a) + g_{\mathrm{eff}}\, u_B \frac{\mathrm{d}}{\mathrm{d}u_0} F(u_0 - u_{\mathrm{LOS}}, a)\right], \\
\psi_{\mathrm{r}} &\simeq \frac{1}{\sqrt{\pi}} \left[F(u_0 - u_{\mathrm{LOS}}, a) - g_{\mathrm{eff}}\, u_B \frac{\mathrm{d}}{\mathrm{d}u_0} F(u_0 - u_{\mathrm{LOS}}, a)\right].
\end{aligned} \tag{8.63}$$

In a more compact form,

$$\phi_\alpha \simeq \frac{1}{\sqrt{\pi}} H(u_0 + \alpha g_{\mathrm{eff}}\, u_B - u_{\mathrm{LOS}}, a), \tag{8.64}$$

and

$$\psi_\alpha \simeq \frac{1}{\sqrt{\pi}} F(u_0 + \alpha g_{\mathrm{eff}}\, u_B - u_{\mathrm{LOS}}, a), \tag{8.65}$$

that is, the profiles are approximated by those of an *effective triplet* whose splitting is that of the normal triplet times the effective Landé factor. The usefulness of the approximation depends on the magnetic field strength, on the central wavelength of the line, and on the particular Zeeman pattern. The first two dependences are obvious: as long as B or λ_0 are small enough, the splittings are small and the first-order approximation may be accurate enough. The influence of the Zeeman pattern is easy to understand as well: one just need compare the approximation accuracy for different Zeeman patterns. Certainly, the effective triplet is an exact description for triplets, no matter what the splitting. Nevertheless, given a field strength and a central wavelength, the approximation is not as good for lines with π and σ components at interlaced positions as for those other having the σ_{b}, π, and σ_{r} spectra well

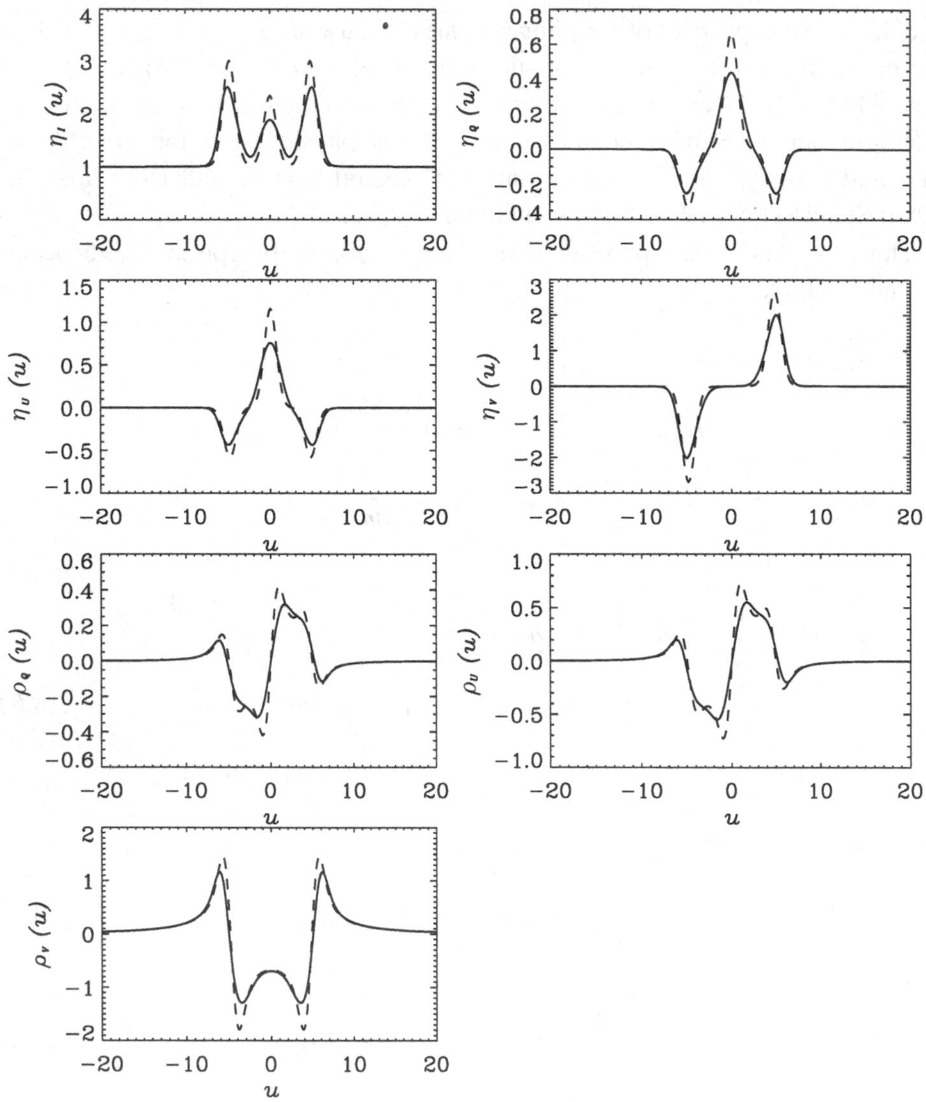

Fig. 8.8. Actual propagation matrix elements of the $^5D_2$—$^7D_3$ transition (solid lines) and those for its effective triplet (dashed lines).

separated from each other. As an example, Figs 8.8 and 8.9 show comparisons between the actual **K** matrix elements and those of the corresponding effective triplet for the $^5D_2$—$^7D_3$ (right, top panel of Fig. 8.5) and $^3P_2$—$^5P_1$ (left, middle panel of Fig. 8.5) transitions. The parameters are the same as for Figs 8.6 and 8.7; specifically, $u_B = 2.4$. While the effective triplet is a reasonable approximation for the first transition, it significantly fails for the second.

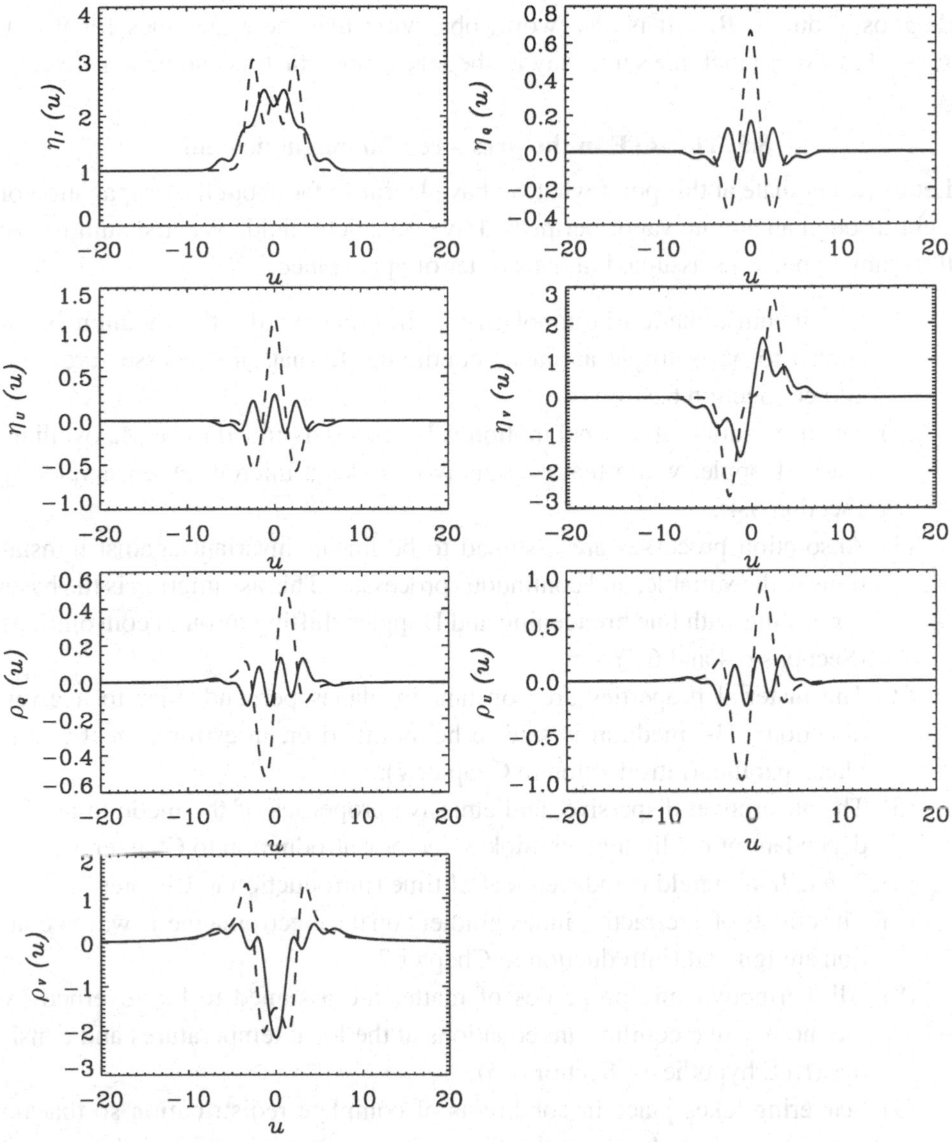

Fig. 8.9. Actual propagation matrix elements of the ${}^3P_2$—${}^5P_1$ transition (solid lines) and those for its effective triplet (dashed lines).

Despite all the drawbacks, the effective Landé factor provides a "thumb-nail" sketch of the magnetic sensitivity of a given spectral line: the larger $g_{\rm eff}$, the larger the splitting. Hence, spectral lines with $g_{\rm eff} = 0$ are unaffected by the presence of a magnetic field. Such lines are often used, for example, to study the thermodynamic and dynamic properties of magnetic atmospheres without any perturbation in the diagnostic due to $\boldsymbol{B}$.† It is also worth observing that there are lines for which $g_{\text{eff}} < 0$, in very much the same way as the level Landé factors can be negative.

8.5 The RTE in the presence of a magnetic field

Let us recapitulate at this point what we have learnt so far about the propagation of light through an atomic vapor permeated by a magnetic field. We first summarize the many hypotheses assumed in their order of appearance:

(1) Continuum radiation is unpolarized. In other words, the medium is assumed to be isotropic as far as continuum-formation processes are concerned (Section 6.3).
(2) The distribution of thermal motion velocities is assumed to be Maxwellian, with a Doppler width that eventually includes a microturbulence velocity (Section 6.4).
(3) Absorption processes are assumed to be linear, invariant against translations of the variable, and continuous processes. This assumption is the basis for dealing with line broadening and Doppler shifting through convolutions (Sections 6.4 and 6.5).
(4) The material properties are constant in planes perpendicular to a given direction. The medium is said to be stratified or, in astrophysical terms, plane–parallel (introduction to Chapter 7).
(5) The absorptive, dispersive, and emissive properties of the medium are independent of the light beam Stokes vector (introduction to Chapter 7).
(6) The radiation field is independent of time (introduction to Chapter 7).
(7) The effects of a refractive index gradient on the electromagnetic wave equation are ignored (introduction to Chapter 7).
(8) All thermodynamic properties of matter are assumed to be governed by thermodynamic equilibrium equations at the local temperatures and densities (LTE hypothesis; Section 7.5).
(9) Scattering takes place in conditions of complete redistribution so that no correlation exists between the frequencies of the incoming and scattered photons (Section 7.5).
(10) All Zeeman sub-levels are equally populated and no coherences exist among them (Section 8.3.3).

With all these assumptions, the transport of radiative energy through the medium is governed by the RTE,

$$\frac{\mathrm{d}\boldsymbol{I}}{\mathrm{d}\tau_{\mathrm{c}}} = \mathbf{K}\,(\boldsymbol{I} - \boldsymbol{S}), \tag{8.66}$$

† The diagnostics provided by the Stokes profiles are discussed in Chapter 10.

where $\boldsymbol{I} = (I, Q, U, V)^{\mathrm{T}}$ is the light beam Stokes vector, τ_{c} is the continuum optical depth defined in Eq. (7.41), $\boldsymbol{S} = (B_{\nu}, 0, 0, 0)^{\mathrm{T}}$ is the source function vector (B_{ν} is the Planck function), and $\mathbf{K}$ is the propagation matrix, whose elements are defined in Eqs (7.44) and (7.45), the normalized absorption and dispersion profiles being given by Eqs (8.50) and (8.51).

It is to be observed that all the thermodynamic, dynamic, magnetic, atomic, and geometric medium properties are included in the propagation matrix and in the source function vector. The optical depth scale (the *natural* length for radiative transfer) is also dependent on the material properties through χ_{cont}. Thermodynamics enters the game through χ_{cont}, η_0 [Eq. (7.40)], and B_{ν}. The latter depends only on the local temperature, T. In their turn, χ_{cont} and η_0 depend on two thermodynamic variables like T and the density, ρ, or the electron pressure, p_{e}, through the Boltzmann and Saha equilibrium population equations, which also involve the abundance of the various ionic species constituting the medium. Thermodynamics also influences the Doppler width and the line damping parameter. The macroscopic motions of the atoms have an influence on the individual Zeeman component profile shapes through Doppler shifts. Atomic parameters like the oscillator strength or quantum numbers are relevant to the evaluation of η_0, ϕ_{α}, and ψ_{α}. Last but not least, the vector magnetic field determines to a large extent the $\mathbf{K}$ matrix element values.

In summary, the radiative transfer equation, being a linear differential equation, contains implicitly all the *non-linear* dependences of the Stokes spectrum on the medium properties. These dependences are non-linear because so are those of τ_{c}, $\mathbf{K}$, and $\boldsymbol{S}$. The intricacy of such dependences may include coupling between dynamical and magnetic parameters with thermodynamic parameters.† The main astrophysical problem, that is, the diagnostics of the properties of the medium based on observations of the Stokes spectrum, turns out to be a formidable problem. Nevertheless, considerable progress has been achieved, as we shall see in the following chapters.

Recommended bibliography

Cohen-Tannoudji, C., Diu, B., and Laloë, F. (1977). *Quantum mechanics* (Hermann and John Wiley & Sons, Inc.). Volume 1, Compl. $\mathrm{D_{VII}}$, Volume 2, Chapter 10.

Cowan, R.D. (1981). *The theory of atomic structure and spectra* (The University of California Press: Berkeley). Chapters, 2, 14, and 17.

Jefferies, J., Lites, B.W., and Skumanich, A.P. (1989). Transfer of line radiation in a magnetic field. *Astrophys. J.* **343**, 920.

Landi Degl'Innocenti, E. (1982). On the effective Landé factor of magnetic lines. *Solar Phys.* **77**, 285.

† Dynamic and magnetic pressure terms as well as magnetic tensions ought to be included in the mechanical equilibrium equations accounting for the stationary state of stellar structures like sunspots.

Landi Degl'Innocenti, E. (1992). Magnetic field measurements, in *Solar observations: techniques and interpretations*. F. Sánchez, M. Collados, and M. Vázquez (eds.) (Cambridge University Press: Cambridge). Sections 4–8.
Landi Degl'Innocenti, E. and Landolfi, M. (2003). *Polarization in spectral lines* (Kluwer Academic Publishers: Dordrecht), in the press.
Rachkovsky, D.N. (1962). Magneto-optical effects in spectral lines of sunspots (in Russian). *Izv. Krymsk. Astrofiz. Obs.* **27**, 148.
Rachkovsky, D.N. (1962). Magnetic rotation effects in spectral lines (in Russian). *Izv. Krymsk. Astrofiz. Obs.* **28**, 259.
Rachkovsky, D.N. (1967). The reduction for anomalous dispersion in the theory of the absorption line formation in a magnetic field (in Russian). *Izv. Krymsk. Astrofiz. Obs.* **37**, 56.
Rees, D.E. (1987). A gentle introduction to polarized radiative transfer, in *Numerical radiative transfer*. W. Kalkofen (ed.) (Cambridge University Press: Cambridge), 213.
Sánchez Almeida, J. (1988). *Estudio de la componente magnética de fáculas y red fotosférica*. PhD Thesis, (Universidad de La Laguna: La Laguna, Spain). Chapter 0.
Solanki, S.K. and Stenflo, J.O. (1985). Models of solar magnetic flux tubes: constraints imposed by Fe I and II lines. *Astron. Astrophys.* **148**, 123.
Stenflo, J.O. (1991). Unified classical theory of line formation in a magnetic field, in *Solar polarimetry*. L.J. November (ed.) (National Solar Observatory/ Sacramento Peak: Sunspot, NM), 416.
Stenflo, J.O. (1994). *Solar magnetic fields. Polarized radiation diagnostics* (Kluwer Academic Publishers: Dordrecht). Chapter 3.
Stenflo, J.O., Harvey, J.W., Brault, J.W., and Solanki, S.K. (1984). Diagnostics of solar magnetic flux tubes using a Fourier transform spectrometer. *Astron. Astrophys.* **131**, 333.
Unno, W. (1956). Line formation of a normal Zeeman triplet. *Pub. Astron. Soc. Japan*, **8**, 108.

9

Solving the radiative transfer equation

...lo que pudiera turbarle en el deliquio sin nombre que gozaba en presencia de Ana, eso aborrecía; lo que pudiera traer una solución al terrible conflicto, cada vez más terrible, de los sentidos enfrentados y de la eternidad pura de su pasión, eso amaba.

—Leopoldo Alas, Clarín, 1885.

Whatever disturbed the nameless rapture that engrossed him in Ana's presence he detested; whatever could bring a solution to the ever more terrible conflict between his constrained senses and the pure eternity of his passion he loved.

With the radiative transfer equation for polarized light to hand, we shall proceed to find solutions and to exploit them both, the equation and its solutions, in order to obtain information about the medium. This chapter is devoted to solutions of the RTE and to the first and simplest diagnostics one can obtain from the observed Stokes profiles. The main emphasis is on concepts rather than numerical details. The latter may be found in the literature (some of the most recent papers are recommended in the bibliography) and in fact are still in continuous evolution and debate. Most of the concepts we describe in this chapter, however, may be said to be well founded nowadays and will help the reader in understanding the topic.

9.1 The model atmosphere

After the summary discussions of Section 8.5, we understand that the medium is usually specified by a number of physical parameters as a function of the geometrical distance, which determines the local value of the optical depth, the propagation matrix, and the source function vector. Using astrophysical terminology, we shall call the set of such parameters the *model atmosphere*, although the medium may be a gas cell in the laboratory. Typically, the model atmosphere contains two thermodynamic variables such as the temperature, T, and the electron pressure, $p_{\rm e}$, the microturbulence velocity, $\xi_{\rm mic}$ (which may not be needed in the laboratory; see Section 6.4), the line-of-sight velocity, $v_{\rm LOS}$ (that is, the component of the material velocity in the propagation direction),† and the three components of the magnetic field vector, B (the strength), θ (the inclination with respect to the propagation direction), and φ (the azimuth with respect to a reference direction, usually, the Stokes Q positive direction). The model atmosphere can then be formally represented by a vector function,

$$\boldsymbol{x}(\tau_{\rm c}) \equiv [T(\tau_{\rm c}), p_{\rm e}(\tau_{\rm c}), \xi_{\rm mic}(\tau_{\rm c}), v_{\rm LOS}(\tau_{\rm c}), B(\tau_{\rm c}), \theta(\tau_{\rm c}), \varphi(\tau_{\rm c}), \xi_{\rm mac}(\tau_{\rm c})], \quad (9.1)$$

where the last parameter is called the *macroturbulence velocity*, a new *ad hoc* astrophysical parameter that accounts for possible motions on scales larger than the mean free path of photons, but that remain spatially unresolved in observations (see Section 9.2.4).

9.2 The formal solution

The propagation matrix, **K**, and the source function vector, $\boldsymbol{S}$, are thus functions of the optical depth through $\boldsymbol{x}(\tau_{\rm c})$; explicitly, $\mathbf{K}\,[\boldsymbol{x}(\tau_{\rm c})]$ and $\boldsymbol{S}\,[\boldsymbol{x}(\tau_{\rm c})]$. Once we know $\boldsymbol{x}(\tau_{\rm c})$, we know $\mathbf{K}(\tau_{\rm c})$ and $\boldsymbol{S}(\tau_{\rm c})$, and we can proceed to solve the RTE,

$$\frac{{\rm d}\boldsymbol{I}}{{\rm d}\tau_{\rm c}} = \mathbf{K}\,(\boldsymbol{I} - \boldsymbol{S}), \quad (9.2)$$

provided we have an initial (or boundary) condition.

9.2.1 Symmetry properties of the solution

Prior to solving the RTE, we can already find a very interesting property of the solution; namely that in the absence of velocity gradients along the optical path the Stokes profiles of every spectral line have definite symmetry properties: Stokes I, Q, and U are even functions of wavelength (or frequency) about λ_0 (or ν_0), and Stokes V is an odd function of wavelength. To understand these properties, let us recall the results on the symmetry of propagation matrix elements of Sections 8.1.1 and 8.4.

According to such results, a change of $(\lambda - \lambda_0) \longrightarrow (\lambda_0 - \lambda)$ transforms the propagation matrix to

$$\mathbf{K}' = \begin{pmatrix} \eta_I & \eta_Q & \eta_U & -\eta_V \\ \eta_Q & \eta_I & \rho_V & \rho_U \\ \eta_U & -\rho_V & \eta_I & -\rho_Q \\ \eta_V & -\rho_U & \rho_Q & \eta_I \end{pmatrix}. \quad (9.3)$$

† The astrophysical convention for velocities defines receding (redshifted) velocities as positive.

Now, if there is no change of the LOS velocity throughout the atmosphere, this is a constant modification of **K** when we are dealing with symmetric wavelengths relative to the central position of the line.† Thus, if $\boldsymbol{I}(0; \lambda - \lambda_0)$ is a solution of the RTE, $\boldsymbol{I}(0; \lambda_0 - \lambda)$ obeys the equation

$$\frac{\mathrm{d}}{\mathrm{d}\tau_c}\boldsymbol{I}(0; \lambda_0 - \lambda) = \mathbf{K}' \, [\boldsymbol{I}(0; \lambda_0 - \lambda) - \boldsymbol{S}] \tag{9.4}$$

and, necessarily,

$$\begin{aligned} I(0; \lambda_0 - \lambda) &= I(0; \lambda - \lambda_0), \\ Q(0; \lambda_0 - \lambda) &= Q(0; \lambda - \lambda_0), \\ U(0; \lambda_0 - \lambda) &= U(0; \lambda - \lambda_0), \\ V(0; \lambda_0 - \lambda) &= -V(0; \lambda - \lambda_0). \end{aligned} \tag{9.5}$$

This is so because, from our hypotheses, the source function is assumed to be constant within the wavelength span of a single spectral line, and the initial (or boundary) condition is assumed to preserve the symmetries.

These remarkable symmetry properties have a further important consequence: in the absence of velocity gradients there is no net circular polarization in a given spectral line; in other words, the integral of Stokes V over the wavelength span, W, of a line is zero:

$$\int_W V(0; u)\, \mathrm{d}u = 0. \tag{9.6}$$

Therefore, as soon as one detects a net circular polarization, a gradient of LOS velocity with optical depth is known to be present in the medium.‡ Practice reveals situations in which the observed Stokes profiles may not follow the symmetry conditions (9.5) but where condition (9.6) is fulfilled. In such cases, one should conclude that there is a lack of spatial resolution, and that the observed signal is a sum of individual signals, each coming from a medium where the Stokes profiles may be different and even shifted in wavelength owing to material velocities, but where no gradient of velocity is present:§ every individual set of Stokes profiles verifies conditions (9.5) so that condition (9.6) is automatically fulfilled; the sum of such shifted profiles may be asymmetric [conditions (9.5) no longer hold], but the integral of Stokes V cannot be different from zero.

† We understand that λ_0 is already shifted by $\Delta\lambda_{\rm LOS}$.

‡ Recall that the possibility of circular polarization of the continuum is already discarded by hypothesis (see Section 8.5).

§ It is important to note that these conclusions hold within the framework of our hypotheses. In particular, a population imbalance between the upper and the lower levels of the transition might produce a net circular polarization.

9.2.2 The evolution operator

Let us consider the homogeneous equation

$$\frac{\mathrm{d}\boldsymbol{I}_h}{\mathrm{d}\tau_\mathrm{c}} = \mathbf{K}\boldsymbol{I}_h, \tag{9.7}$$

where $\boldsymbol{I}_h$ represents a solution of this homogeneous equation. Let $\mathbf{O}(\tau_\mathrm{c}, \tau_\mathrm{c}')$ be a *linear* operator which gives the transformation of the homogeneous solution between the two points at optical depths τ_c' and τ_c:

$$\boldsymbol{I}_h(\tau_\mathrm{c}) \equiv \mathbf{O}(\tau_\mathrm{c}, \tau_\mathrm{c}')\boldsymbol{I}_h(\tau_\mathrm{c}'). \tag{9.8}$$

$\mathbf{O}(\tau_\mathrm{c}, \tau_\mathrm{c}')$ is known as the *evolution operator* and obviously fulfils

$$\mathbf{O}(\tau_\mathrm{c}, \tau_\mathrm{c}) = \mathbb{1} \tag{9.9}$$

and

$$\mathbf{O}(\tau_\mathrm{c}, \tau_\mathrm{c}'') = \mathbf{O}(\tau_\mathrm{c}, \tau_\mathrm{c}')\mathbf{O}(\tau_\mathrm{c}', \tau_\mathrm{c}''). \tag{9.10}$$

Hence, the variation of $\boldsymbol{I}_h$ when τ_c varies is given by

$$\frac{\mathrm{d}\boldsymbol{I}_h(\tau_\mathrm{c})}{\mathrm{d}\tau_\mathrm{c}} = \frac{\mathrm{d}\mathbf{O}(\tau_\mathrm{c}, \tau_\mathrm{c}')}{\mathrm{d}\tau_\mathrm{c}}\boldsymbol{I}_h(\tau_\mathrm{c}'). \tag{9.11}$$

On the other hand, Eqs (9.7) and (9.8) imply that

$$\frac{\mathrm{d}\boldsymbol{I}_h(\tau_\mathrm{c})}{\mathrm{d}\tau_\mathrm{c}} = \mathbf{K}(\tau_\mathrm{c})\,\mathbf{O}(\tau_\mathrm{c}, \tau_\mathrm{c}')\boldsymbol{I}_h(\tau_\mathrm{c}'). \tag{9.12}$$

Now, comparing Eqs (9.11) and (9.12), we conclude that the evolution operator must verify the following differential equation:

$$\frac{\mathrm{d}\mathbf{O}(\tau_\mathrm{c}, \tau_\mathrm{c}')}{\mathrm{d}\tau_\mathrm{c}} = \mathbf{K}(\tau_\mathrm{c})\,\mathbf{O}(\tau_\mathrm{c}, \tau_\mathrm{c}'). \tag{9.13}$$

Since $\boldsymbol{I}_h(\tau_\mathrm{c})$ is independent of τ_c', similar steps to those taken to obtain Eqs (9.11) and (9.12), but now accounting for variations with respect to τ_c', give

$$\frac{\mathrm{d}\mathbf{O}(\tau_\mathrm{c}, \tau_\mathrm{c}')}{\mathrm{d}\tau_\mathrm{c}'} = -\mathbf{O}(\tau_\mathrm{c}, \tau_\mathrm{c}')\,\mathbf{K}(\tau_\mathrm{c}'). \tag{9.14}$$

9.2.3 Solving the inhomogeneous equation

Now, we can use the evolution operator as an "integrating factor". Let us multiply the RTE [Eq. (9.2)] by $\mathbf{O}(\tau_\mathrm{c}', \tau_\mathrm{c})$ to obtain

$$\mathbf{O}(\tau_\mathrm{c}', \tau_\mathrm{c})\frac{\mathrm{d}\boldsymbol{I}}{\mathrm{d}\tau_\mathrm{c}} = \mathbf{O}(\tau_\mathrm{c}', \tau_\mathrm{c})\,\mathbf{K}\,(\boldsymbol{I} - \boldsymbol{S}). \tag{9.15}$$

The left-hand side of Eq. (9.15) is

$$\mathbf{O}(\tau_c', \tau_c)\frac{\mathrm{d}\boldsymbol{I}}{\mathrm{d}\tau_c} = \frac{\mathrm{d}}{\mathrm{d}\tau_c}\left[\mathbf{O}(\tau_c', \tau_c)\,\boldsymbol{I}(\tau_c)\right] - \frac{\mathrm{d}\mathbf{O}(\tau_c', \tau_c)}{\mathrm{d}\tau_c}\boldsymbol{I}(\tau_c), \tag{9.16}$$

and using now Eq. (9.14), Eqs (9.15) and (9.16) give

$$\frac{\mathrm{d}}{\mathrm{d}\tau_c}\left[\mathbf{O}(\tau_c', \tau_c)\,\boldsymbol{I}(\tau_c)\right] = -\mathbf{O}(\tau_c', \tau_c)\,\mathbf{K}(\tau_c)\,\boldsymbol{S}(\tau_c), \tag{9.17}$$

which, after integration with respect to optical depth between τ_0 and τ_1, results in

$$\boldsymbol{I}(\tau_1) = \mathbf{O}(\tau_1, \tau_0)\,\boldsymbol{I}(\tau_0) - \int_{\tau_0}^{\tau_1} \mathbf{O}(\tau_1, \tau_c)\,\mathbf{K}(\tau_c)\,\boldsymbol{S}(\tau_c)\,\mathrm{d}\tau_c, \tag{9.18}$$

where we have made use of Eq. (9.9). The first term on the right-hand side of Eq. (9.18) gives us the Stokes parameters of the light that has streamed through the medium (between τ_0 and τ_1) as if the medium were not emitting light (the homogeneous solution); $\mathbf{O}$ describes such an evolution. The second term of the equation provides the contribution from emission to the final Stokes parameters. Note that emission is accounted for by $\mathbf{K}\,\boldsymbol{S}$. That emitted light in τ_c "evolves" up to τ_1 and is then "added". A final remark is in order at this point: our adding of the Stokes parameters implies that we are assuming the absence of coherences between the light beams emitted at different points in the medium. Should these coherences exist, the Stokes formalism would be inappropriate (see Section 3.4).

If we assume a semi-infinite medium, as is customary when considering stellar atmospheres, we accept that the medium has an open external boundary where the origin of optical depths is located ($\tau_1 = 0$; the observer's position, in fact) and an internal boundary where the material is so optically thick that the mathematical limit $\tau_0 \longrightarrow \infty$ makes sense. For this type of medium, one often assumes that, in the limit, no photon at the internal boundary may reach the external boundary , i.e.,

$$\lim_{\tau_0 \longrightarrow \infty} \mathbf{O}(0, \tau_0)\,\boldsymbol{I}(\tau_0) = 0. \tag{9.19}$$

We can then write the formal solution of the RTE as†

$$\boldsymbol{I}(0) = \int_0^{\infty} \mathbf{O}(0, \tau_c)\,\mathbf{K}(\tau_c)\,\boldsymbol{S}(\tau_c)\,\mathrm{d}\tau_c. \tag{9.20}$$

9.2.4 The action of macroturbulence on the Stokes profiles

Observed Stokes profiles are often wider than synthetic profiles of the same equivalent width, that is, profiles that withdraw the same amount of electromagnetic energy from the continuum. This effect can be attributed to the presence of turbulent

† This formal solution of the RTE for polarized light was proposed for the first time by Landi Degl'Innocenti and Landi Degl'Innocenti (1985); see also Landi Degl'Innocenti (1987).

motions that remain unresolved within the spatial resolution element. The distribution of material velocities certainly has an influence on the Stokes profiles. Such an influence may be *tailored* at will if one assumes that the turbulence velocities vary with depth through the atmosphere. This assumption may be of help to those analytical techniques that are otherwise unable to reproduce the observed profiles. However, macroturbulence (that is, turbulence on a scale larger than the mean free path of photons) is mostly called on in an *ad hoc* manner, rather than on the basis of actual physical reasoning. Thus, we shall make the conservative assumption‡ that macroturbulence is constant with depth and therefore acts as a convolution of the Stokes profiles with a Gaussian distribution of velocities of width ξ_{mac}:

$$G_{\mathrm{mac}} \equiv \frac{1}{\sqrt{\pi}\,\xi_{\mathrm{mac}}}\,\mathrm{e}^{-\left(v^2/\xi_{\mathrm{mac}}^2\right)}. \tag{9.21}$$

Note that G_{mac} is normalized in area, so that convolution will not change the equivalent width of Stokes I. In other words, energy is preserved during the process. It is convenient to consider G_{mac} as a distribution of wavelength displacements. Following the same steps as in Section 6.4, the macroturbulence distribution is

$$G_{\mathrm{mac}} = \frac{1}{\sqrt{\pi}\,\sigma_{\mathrm{mac}}}\,\mathrm{e}^{-\left[(\lambda-\lambda_0)^2/\sigma_{\mathrm{mac}}^2\right]}, \tag{9.22}$$

where

$$\sigma_{\mathrm{mac}} \equiv \frac{\lambda_0 \xi_{\mathrm{mac}}}{c}. \tag{9.23}$$

This distribution is convolved with the Stokes spectrum to get the observed profiles

$$I_{\mathrm{obs},i} = I_i * G_{\mathrm{mac}}, \tag{9.24}$$

where $i = 1, 2, 3, 4$ runs for all four Stokes parameters. Figure 9.1 illustrates the effect of macroturbulence on the profiles. The Fe I line at 630.25 nm is synthesized in the Harvard–Smithsonian Reference Atmosphere (HSRA; Gingerich *et al.*, 1971) to which a constant magnetic field vector has been added, without macroturbulence (dashed lines) and with $\xi_{\mathrm{mac}} = 2\ \mathrm{km\,s^{-1}}$ (solid lines).

9.3 Actual (numerical) solutions of the RTE

The solution (9.20) of the RTE is called *formal* because it is indeed not a *real* solution as long as the evolution operator is not known. Unfortunately, despite their fairly simple appearance, Eqs (9.13) and (9.14) have no easy analytical solution in

‡ It is a conservative hypothesis in order not to enlarge artificially the number of free parameters in numerical inferences (see Chapter 11).

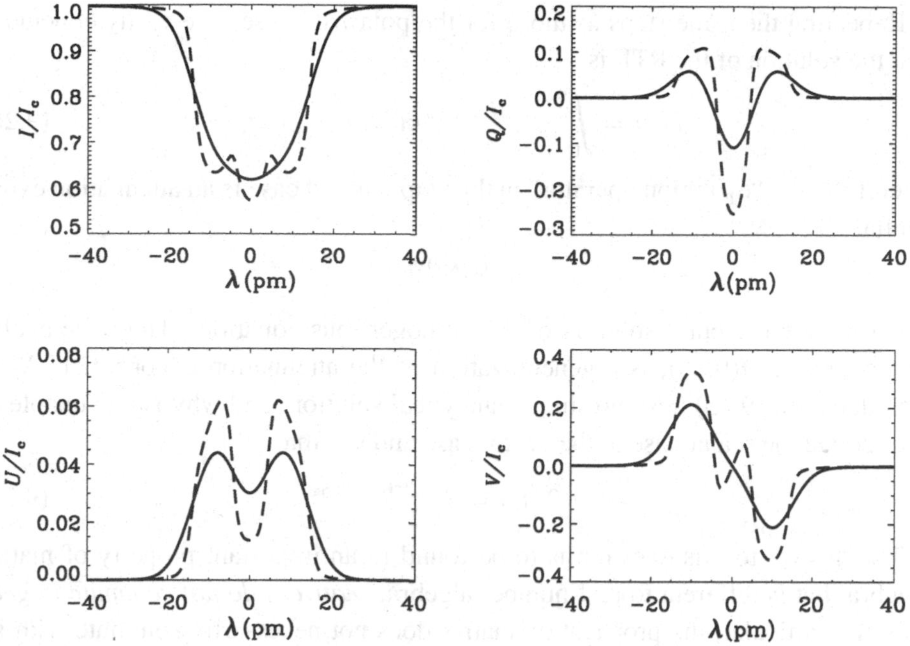

Fig. 9.1. Stokes profiles of the Fe I line at 630.25 nm in the HSRA model with a constant magnetic field vector of $B = 2000$ G, $\theta = 45°$, and $\varphi = 10°$, without macroturbulence (dashed lines) and with a macroturbulence velocity of 2 km s^{-1}(solid lines). The Stokes profiles are normalized by the continuum intensity, I_c.

general. Only in particular instances can these equations be integrated analytically. In most instances, only numerical approaches to the evolution operator can be found. We shall understand this by confronting the RTE for anisotropic media (hence, for polarized light) with its particular case for isotropic media (hence, where polarization can be forgotten and a scalar equation for the intensity can be considered alone).† After the comparison we shall better realize one of the most important characteristics of the polarized radiative transport, namely the needs for matrix algebra and the difficulties that this algebra entails.

Consider an isotropic medium with a line-to-continuum absorption coefficient ratio η_0 and an absorption profile ϕ (both invariable for all directions) and a source function that, in the LTE approximation, is given by the Planck function at the local temperature. Through such a medium light propagates according to the equation

$$\frac{\mathrm{d}I}{\mathrm{d}\tau_c} = k(\tau_c)\ [I(\tau_c) - B_\nu(\tau_c)], \tag{9.25}$$

where here $k \equiv 1 + \eta_0\phi$ [see Eq. (7.46)].

† Recall, however, the discussion of Section 7.7: polarization can be forgotten only when light is initially unpolarized.

Repeating the same steps as those for the polarized case, one easily concludes that the solution of the RTE is

$$I(0) = \int_0^\infty \mathrm{e}^{-\int_0^{\tau_c} k(t)\,\mathrm{d}t} k(\tau_c)\, B_\nu(\tau_c)\, \mathrm{d}\tau_c. \tag{9.26}$$

Therefore, the "evolution operator" in this unpolarized case is an attenuation exponential

$$\mathrm{e}^{-\int_0^{\tau_c} k(t)\,\mathrm{d}t},$$

which is nothing but a solution of the homogeneous equation. Thus, the evolution operator, $\mathbf{O}(0, \tau_c)$, is a generalization of the attenuation exponential. Why then does Eq. (9.13) have no simple analytical solution, and why is one unable to extrapolate the scalar case to the vector case and say that

$$\mathbf{O}(0, \tau_c) = \mathrm{e}^{-\int_0^{\tau_c} \mathbf{K}(t)\,\mathrm{d}t}? \tag{9.27}$$

The answer to this question is to be found in an important property of matrix algebra that is different to real-number algebra: *matrices do not commute in general*. In particular, the propagation matrix does not necessarily commute with its integral over the optical path,

$$\left[\mathbf{K}, \int_0^{\tau_c} \mathbf{K}(t)\mathrm{d}t\right] \neq \mathbf{0}, \tag{9.28}$$

where the square brackets indicate the commutation operator

$$([\mathbf{A}, \mathbf{B}] \equiv \mathbf{AB} - \mathbf{BA}).$$

Therefore, in general,

$$\frac{\mathrm{d}}{\mathrm{d}\tau_c}\mathrm{e}^{-\int_0^{\tau_c} \mathbf{K}(t)\,\mathrm{d}t} \neq -\mathrm{e}^{-\int_0^{\tau_c} \mathbf{K}(t)\,\mathrm{d}t}\,\mathbf{K}(\tau_c), \tag{9.29}$$

so that Eq. (9.14) does not hold. This is so because the exponential of a matrix is to be understood as the result of a Taylor expansion:

$$\mathrm{e}^{\mathbf{A}} \equiv \mathbb{1} + \mathbf{A} + \frac{1}{2!}\mathbf{A}^2 + \frac{1}{3!}\mathbf{A}^3 + \ldots. \tag{9.30}$$

Hence, if we call

$$\mathbf{L}(\tau_c) \equiv \int_0^{\tau_c} \mathbf{K}(t)\,\mathrm{d}t, \tag{9.31}$$

then, in general,

$$\frac{\mathrm{d}}{\mathrm{d}\tau_c}\mathbf{L}^n \neq n\,\mathbf{K}\,\mathbf{L}^{n-1} \tag{9.32}$$

because if $[\mathbf{K}, \mathbf{L}] \neq \mathbf{0}$ then $[\mathbf{K}, \mathbf{L}^m] \neq \mathbf{0}$.

In summary, only when the commutation condition,

$$[\mathbf{K}, \mathbf{L}] = \mathbf{KL} - \mathbf{LK} = \mathbf{0}, \tag{9.33}$$

holds is the matrix exponential of Eq. (9.27) a valid expression for the evolution operator. In such a case, the formal solution of the RTE becomes

$$\boldsymbol{I}(0) = \int_0^{\infty} \mathrm{e}^{-\int_0^{\tau_\mathrm{c}} \mathbf{K}(t)\,\mathrm{d}t}\, \mathbf{K}(\tau_\mathrm{c})\, \boldsymbol{S}(\tau_\mathrm{c})\, \mathrm{d}\tau_\mathrm{c}. \tag{9.34}$$

Otherwise, numerical approaches must be found: the most general expressions which have been obtained so far involve infinite series, which are not very easy to handle in practice (see Landi Degl'Innocenti, 1987; López Ariste and Semel, 1999). A specific and simple case for which the commutation condition (9.33) holds is that of a medium with a constant propagation matrix or, more generally, when $\mathbf{K}(\tau_\mathrm{c}) = f(\tau_\mathrm{c})\,\mathbf{K}_0$, where $\mathbf{K}_0$ is constant and f is a scalar function of the optical depth. This case will be further constrained later (Section 9.4.1) in order to obtain an analytical solution to the RTE that may help in gaining an insight into the problem.

We shall not enter here into the details of the various numerical solutions and refer the interested reader to Landi Degl'Innocenti (1987), Bellot Rubio *et al.* (1998), and Semel and López Ariste (1999). For our purposes, we shall assume that any of these numerical solutions provides a "good" evolution operator and, thus, our discussions can circumscribe the formal solution (9.20) just as if it were a true analytical solution. Therefore, from now on we shall understand that, given the stratification of atmospheric parameters, an (accurate enough) solution of the RTE can be calculated. An example of such a solution, i.e., the spectrum of all four Stokes parameters, for the Fe I line at 630.25 nm is shown in Fig. 9.2. The model atmosphere, that is, the stratification of physical quantities, corresponding to that solution is shown in Fig. 9.3. The temperature corresponds to a mean penumbral model by del Toro Iniesta *et al.* (1994); the electron pressure stratification is compatible with a gas pressure stratification obeying the hydrostatic equilibrium equation; the functional shape of the line-of-sight velocity of the material and of the three components of the magnetic field are not meant to resemble any realistic stratification; the microturbulence velocity is 0.6 $\mathrm{km\,s^{-1}}$and the macroturbulence velocity is 0.75 $\mathrm{km\,s^{-1}}$. The profiles are normalized to the continuum of the quiet Sun (HSRA).

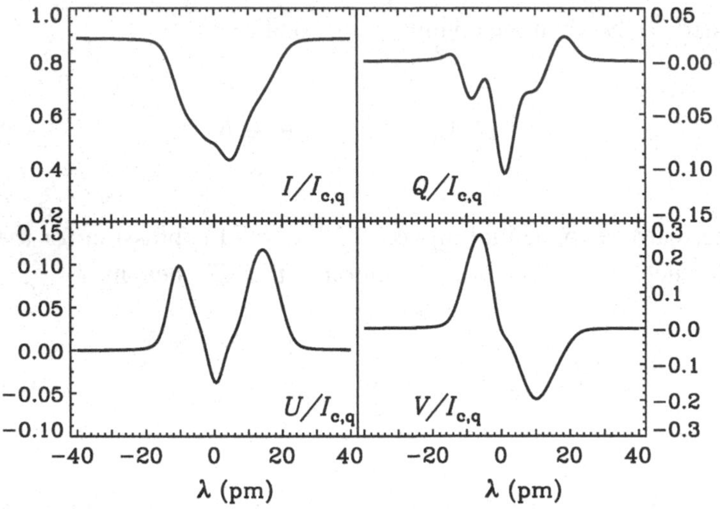

Fig. 9.2. Stokes profiles of the Fe I line at 630.25 nm in the model atmosphere shown in Fig. 9.3.

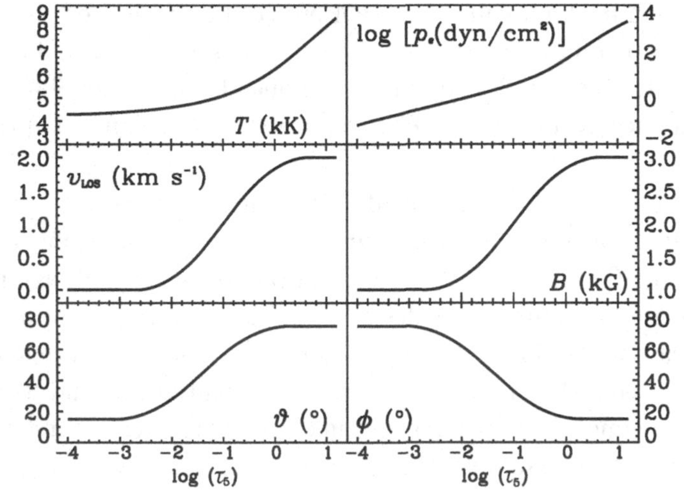

Fig. 9.3. Model atmosphere where the Stokes profiles of Fig. 9.2 have been synthesized.

9.4 Simple solutions of the RTE

Let us assume that the stratification of physical quantities through the atmosphere is such that the propagation matrix can be written as

$$\mathbf{K}(\tau_c) = \mathbf{K}_0\, f(\tau_c), \tag{9.35}$$

where $\mathbf{K}_0$ is constant (independent of the optical depth) and f is any scalar function of τ_c. Under this assumption, the integral, $\mathbf{L}$, over the optical path of the propagation matrix is

$$\mathbf{L}(\tau_c) = \mathbf{K}_0 \int_0^{\tau_c} f(t)\,\mathrm{d}t, \tag{9.36}$$

and the commutator between $\mathbf{K}$ and $\mathbf{L}$ becomes

$$[\mathbf{K}, \mathbf{L}] = \mathbf{K}_0 \left[f(\tau_c), \int_0^{\tau_c} f(t)\,\mathrm{d}t \right] = \mathbf{0}, \tag{9.37}$$

because any two real numbers commute.

By simply applying the definition (9.30) for the exponential of a matrix and the commutation condition (9.37), one can easily check that now

$$\frac{\mathrm{d}}{\mathrm{d}\tau_c} \mathrm{e}^{-\mathbf{L}} = -\mathrm{e}^{-\mathbf{L}} \mathbf{K} \tag{9.38}$$

and, hence, that the analytic expression (9.27) for the evolution operator holds.

9.4.1 The Milne–Eddington atmosphere

Consider now the simplest case among those for which Eq. (9.35) is valid, namely, that medium where

$$\mathbf{K}(\tau_c) = \mathbf{K}_0. \tag{9.39}$$

In such a case, the evolution operator simply reads

$$\mathbf{O}(0, \tau_c) = \mathrm{e}^{-\mathbf{K}_0 \tau_c}. \tag{9.40}$$

Also, assume that the source function vector, $\boldsymbol{S}$, depends linearly on optical depth,

$$\boldsymbol{S} \equiv \boldsymbol{S}_0 + \boldsymbol{S}_1 \tau_c = (S_0 + S_1 \tau_c)(1, 0, 0, 0)^{\mathrm{T}}, \tag{9.41}$$

since we already know (see Section 7.5) that the source function is proportional to $(1, 0, 0, 0)^{\mathrm{T}}$ within the framework of our hypotheses.

A medium satisfying Eqs (9.39) and (9.41) is called in the astrophysical literature a *Milne–Eddington atmosphere*. In this type of atmosphere, the propagation of polarized light is such that the formal solution becomes

$$\boldsymbol{I}(0) = \int_0^{\infty} \mathrm{e}^{-\mathbf{K}_0 \tau_c} \mathbf{K}_0 (\boldsymbol{S}_0 + \boldsymbol{S}_1 \tau_c)\,\mathrm{d}\tau_c, \tag{9.42}$$

which can be integrated analytically by parts to yield

$$\boldsymbol{I}(0) = \boldsymbol{S}_0 + \mathbf{K}_0^{-1} \boldsymbol{S}_1. \tag{9.43}$$

Now, since the only non-zero element of both $\boldsymbol{S}_0$ and $\boldsymbol{S}_1$ is the first one, we only need to calculate the first column of the inverse $\mathbf{K}_0^{-1}$ of the propagation matrix. The explicit expression of all four Stokes profiles is then

$$
\begin{aligned}
I(0) &= S_0 + \Delta^{-1}\eta_I(\eta_I^2 + \rho_Q^2 + \rho_U^2 + \rho_V^2)S_1, \\
Q(0) &= -\Delta^{-1}\left[\eta_I^2\eta_Q + \eta_I(\eta_V\rho_U - \eta_U\rho_V) + \rho_Q(\eta_Q\rho_Q + \eta_U\rho_U + \eta_V\rho_V)\right] S_1, \\
U(0) &= -\Delta^{-1}\left[\eta_I^2\eta_U + \eta_I(\eta_Q\rho_V - \eta_V\rho_Q) + \rho_U(\eta_Q\rho_Q + \eta_U\rho_U + \eta_V\rho_V)\right] S_1, \\
V(0) &= -\Delta^{-1}\left[\eta_I^2\eta_V + \eta_I(\eta_U\rho_Q - \eta_Q\rho_U) + \rho_V(\eta_Q\rho_Q + \eta_U\rho_U + \eta_V\rho_V)\right] S_1,
\end{aligned}
\tag{9.44}
$$

where Δ is the determinant of the propagation matrix; explicitly,

$$\Delta = \eta_I^2(\eta_I^2 - \eta_Q^2 - \eta_U^2 - \eta_V^2 + \rho_Q^2 + \rho_U^2 + \rho_V^2) - (\eta_Q\rho_Q + \eta_U\rho_U + \eta_V\rho_V)^2. \tag{9.45}$$

The solution (9.44) of the RTE in a Milne–Eddington atmosphere is known as the Unno–Rachkovsky solution and is very useful because of its analytical character. Just by changing a few parameters we can get an approximate idea of how a given spectral line is formed in different atmospheres or how various lines are formed in a given atmosphere. Although the model may not be realistic (fairly large variations with depth are expected in actual atmospheres), the Unno–Rachkovsky solution is very helpful.

Let us recapitulate and find out which parameters must be kept constant throughout the medium for the propagation matrix to remain independent of depth. This is very simple since we just have to look at Eqs (8.14), (8.15), (8.50), and (8.51). There, the explicit expressions of the **K** matrix elements are shown as a function of the line-to-continuum absorption coefficient ratio, η_0, the Doppler width of the line, $\Delta\lambda_{\rm D}$, the damping parameter, a (in units of $\Delta\lambda_{\rm D}$), the central wavelength of the line, and the three components (B, θ, and φ) of the magnetic field vector. Hence, the whole model is specified by these seven parameters plus the two parameters describing the source function, S_0 and S_1.

An example of the Unno–Rachkovsky solution is shown in Fig. 9.4, where the Milne–Eddington parameters are $\eta_0 = 10$, $a = 0.05$, $u_B = 2.4$, $\theta = \pi/4$, $\varphi = \pi/6$, and $S_1/S_0 = 4$. The profiles are normalized to their local continuum.

9.4.2 Longitudinal magnetic field

Consider a magnetic atmosphere that is observed in the (constant) direction of $\boldsymbol{B}$. Since it is the magnetic field vector that establishes an optical anisotropy in the medium, we are dealing with a case in which the light is propagating along the

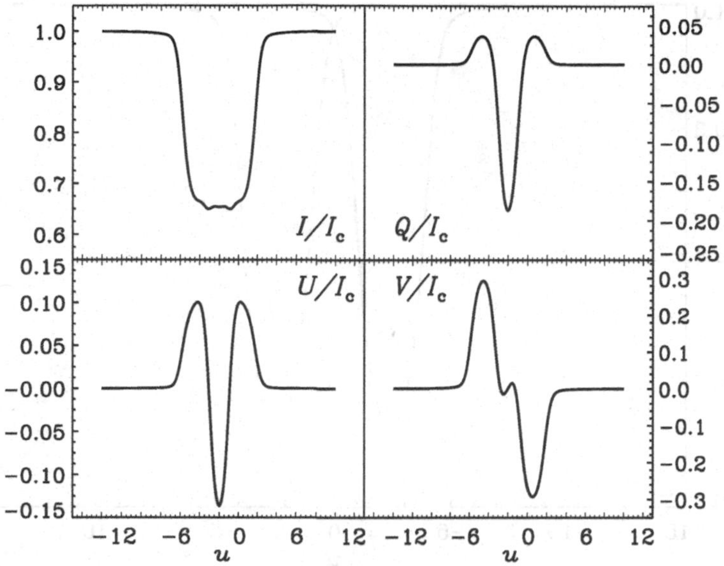

Fig. 9.4. Stokes profiles of a Zeeman triplet in a Milne–Eddington atmosphere of parameters $\eta_0 = 10$, $a = 0.05$, $u_B = 2.4$, $\theta = \pi/4$, $\varphi = \pi/6$, and $S_1/S_0 = 4$. The abscissae are expressed in the reduced wavelength variable defined in Eq. (6.50) [or the reduced frequency of Eq. (6.49)].

optical axis and, therefore, the RTE is simplified (Section 7.8) to

$$\frac{\mathrm{d}}{\mathrm{d}\tau_c}(I \pm V) = (\eta_I \pm \eta_V)(I \pm V - B_\nu), \tag{9.46}$$

where we have explicitly made the Planck function the source function.

We have now two *uncoupled* scalar equations that are formally identical to that for Stokes I in the same atmosphere but in the absence of a magnetic field. In such a case, we can write

$$\frac{\mathrm{d}I}{\mathrm{d}\tau_c} = (1 + \eta_0\phi)(I - B_\nu) \tag{9.47}$$

independently of the other three Stokes parameters. In Eq. (9.47), η_0 is, as usual, the line-to-continuum absorption coefficient ratio and ϕ the (isotropic) absorption profile. Now, when propagation is in the magnetic field direction, according to Eqs (7.50) and (7.52),

$$\eta_I \pm \eta_V = 1 + \eta_0\phi_{\mp 1}, \tag{9.48}$$

where an explicit distinction is made between right-handed and left-handed polarized light.

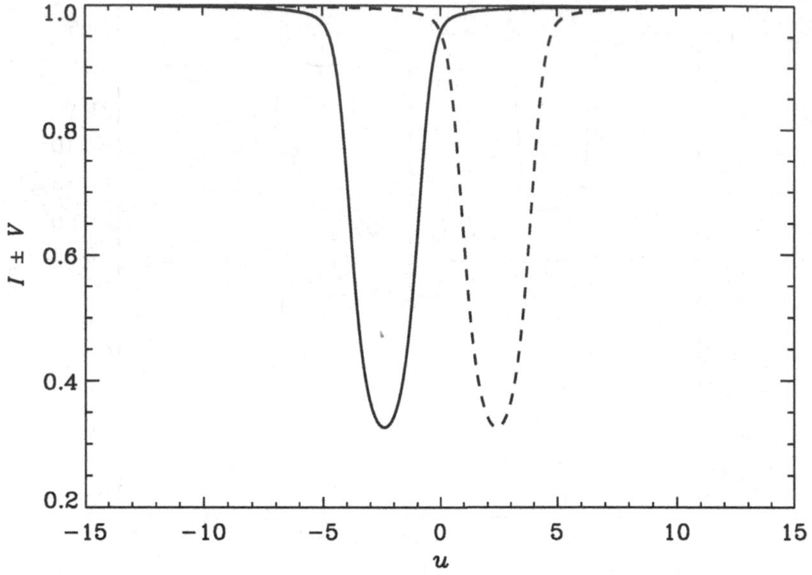

Fig. 9.5. $I \pm V$ profiles of a Zeeman triplet in a Milne–Eddington atmosphere.

Therefore, if in the absence of a magnetic field the solution of the RTE is

$$I(0) = \int_0^\infty \mathrm{e}^{-\int_0^{\tau_c}(1+\eta_0\phi)\mathrm{d}\tau'}(1+\eta_0\phi)\, B_\nu \, \mathrm{d}\tau_c, \tag{9.49}$$

in the case of a longitudinal magnetic field we have

$$(I \pm V)(0) = \int_0^\infty \mathrm{e}^{-\int_0^{\tau_c}(1+\eta_0\phi_{\mp 1})\mathrm{d}\tau'}(1+\eta_0\phi_{\mp 1})\, B_\nu \, \mathrm{d}\tau_c. \tag{9.50}$$

In the particular case of a Zeeman triplet, we already know that the sole difference between ϕ and ϕ_{-1} (or ϕ_{+1}) is a wavelength shift: the absorption profiles are identical (although scaled) but shifted to the blue or to the red depending on the handedness of the polarization (Section 8.3.2). Therefore, the solutions for $I \pm V$ in the longitudinal case are the same as those for I in the absence of $\boldsymbol{B}$, but shifted in wavelength, according to the Zeeman splittings (see Fig. 9.5).

9.5 Simple diagnostics

A few simple physical conclusions can be drawn from a glance at the Stokes profiles. A comparison of Figs 9.2 and 9.4 reveals several highly significant differences. First, the profiles of Fig. 9.2 are very asymmetric whereas those of Fig. 9.4 have definite symmetry properties: Stokes I, Q, and U are symmetric and Stokes

V is antisymmetric with respect to the central wavelength of the line. Second, the Stokes profiles of Fig. 9.4 have a continuum intensity equal to one whereas those of Fig. 9.2 do not. Third, the Unno–Rachkovsky profiles are all shifted by $2\Delta\lambda_{\rm D}$ with respect to the original position of the line in the absence of a velocity field.

These three differences have simple explanations which in fact offer valuable information about the medium in which the Stokes profiles were formed. The difference in symmetry properties and, most importantly, the net circular polarization of penumbral profiles respond to the presence of a velocity gradient with depth in the penumbral model, whereas the Milne–Eddington one has a velocity along the LOS that is constant with optical depth (see Section 9.2.1). Indeed, the shift of the Unno–Rachkovsky profiles implies a mean or bulk velocity (along the LOS) of the material approaching the observer: the shift is bluewards.

The continuum intensity of the penumbral profiles tells us that our spectral line is formed in a medium which is cooler than the quiet Sun. Of course, a quantification of such cooling requires further calculation but the information is there in the profiles. Unfortunately, the Unno–Rachkovsky profiles have been normalized to their local continuum so that any information of relative cooling or heating is lost. We have stressed this circumstance because it is relatively customary among observers to normalize the observed Stokes profiles to their local continuum. Such a practice should be avoided in order to exploit to the maximum the information about the medium carried by the Stokes profiles.

These are only a few very simple examples of the diagnostics that can be carried out on the Stokes profiles. Of course, quantification of all the physical parameters characterizing the medium through which light is traveling requires the more detailed analysis and calculations that are discussed in the following chapters.

Recommended bibliography

Auer, L.H. and House, J.N. (1978). The origin of broad-band circular polarization in sunspots. *Astron. Astrophys.* **64**, 67.

Bellot Rubio, L.R., Ruiz Cobo, B., and Collados, M. (1998). An Hermitian method for the solution of polarized radiative transfer problems. *Astrophys. J.* **506**, 805.

Gingerich, O., Noyes, R.W., Kalkofen, W., and Cuny, Y. (1971). The Harvard-Smithsonian reference atmosphere. *Solar Phys.* **18**, 347.

Landi Degl'Innocenti, E. (1987). Transfer of polarized radiation, using **4** × **4** matrices, in *Numerical radiative transfer*. W. Kalkofen (ed.) (Cambridge University Press: Cambridge), p. 265.

Landi Degl'Innocenti, E. and Landi Degl'Innocenti, M. (1981). Radiative transfer for polarized radiation: symmetry properties and geometrical interpretation. *Il Nuovo Cimento* **62B**, 1.

Landi Degl'Innocenti, E. and Landi Degl'Innocenti, M. (1985). On the solution of the radiative transfer equations for polarized radiation. *Solar Phys.* **97**, 239.

López Ariste, A. (1999). *La spectropolarimétrie en astrophysique. Application au diagnostic des champs magnétiques solaires et stellaires*. PhD Thesis (Université Paris 7: Paris). Chapter 6.

López Ariste, A. and Semel, M. (1999). Analytical solution of the radiative transfer equation for polarized light. *Astron. Astrophys.* **350**, 1089.

Rachkovsky, D.N. (1962). Magneto-optical effects in spectral lines of sunspots (in Russian). *Izv. Krymsk. Astrofiz. Obs.* **27**, 148.

Rachkovsky, D.N. (1962). Magnetic rotation effects in spectral lines (in Russian). *Izv. Krymsk. Astrofiz. Obs.* **28**, 259.

Rachkovsky, D.N. (1967). The reduction for anomalous dispersion in the theory of the absorption line formation in a magnetic field (in Russian). *Izv. Krymsk. Astrofiz. Obs.* **37**, 56.

Rees, D.E., Murphy, G.A., and Durrant, C.J. (1989). Stokes profile analysis and vector magnetic fields. II. Formal numerical solutions of the Stokes transfer equations. *Astrophys. J.* **339**, 1093.

Ruiz Cobo, B. (1992). *Inversión de la ecuación de transporte radiativo*. PhD Thesis, (Universidad de La Laguna: La Laguna, Spain). Chapters 2, 3.

Semel, M. and López Ariste, A. (1999). Integration of the radiative transfer equation for polarized light: the exponential solution. *Astron. Astrophys.* **342**, 201.

del Toro Iniesta, J.C., Tarbell, T.D., and Ruiz Cobo, B. (1994) On the temperature and velocity through the photosphere of a sunspot penumbra. *Astrophys. J.* **436**, 400.

Unno, W. (1956). Line formation of a normal Zeeman triplet. *Pub. Astron. Soc. Japan*, **8**, 108.

10

Stokes spectrum diagnostics

Dixo mio Çid: "yo desto so pagado;
"quando agora son buenos, adelant serán preçiados."

—*Anonymous, approx. 1140.*

'With that I am well paid,' said the Cid;
'Those that are now worthy, shall henceforth be rewarded.'

The main problem in astrophysics is that of inferring the physical properties of the medium from the observables: the Stokes spectrum. Unfortunately, no *in situ* measurements can be made of the temperatures, densities, velocities, magnetic fields, and other physical quantities to probe the astronomical object, or at least that portion of the astronomical object where photons come from. Astrophysical measurements are of the physical properties of the (polarized) radiation, not of the celestial object itself. From these measurements, and with the help of some known physics, the astronomer is challenged to infer the properties of the medium that light has passed through. Certainly, we speak loosely when we use the same word *measurement* for both the process of characterizing light and that of interpreting the observed Stokes spectrum in terms of the medium properties: calibration is neither as easy nor as accurate as in laboratory measurements. The only available "meter" is the RTE, which contains the relationship between the observable (the Stokes spectrum) and the unknowns (the medium physical quantities). More specifically, the link between the medium and the observable lies in the coefficients of the RTE, namely, the propagation matrix and the source function vector. Since simplifying assumptions and models are always necessary in order to particularize both $\mathbf{K}$ and $\boldsymbol{S}$, the results of the inference process depend on these assumptions and models. For instance, one may be interested in *measuring* (inferring) the magnetic field strength of a given region of the Sun, under the assumption that the field does not vary with height. The result will certainly not be the same as that arising from the assumption that the magnetic field strength varies linearly with height. Whatever the hypotheses are, the crucial problem is to determine the dependence of the observable on the various physical quantities characterizing the medium; in other words, to determine the diagnostic capabilities of the particular observable.

Measurements are sometimes not just a resolved portion of the spectrum but the result of some operator acting on it. For instance, some magnetographs measure the integral of Stokes V over a narrow region, and tunable birefringent filters provide narrow-band images across the wavelength span of a given line. We also need to know the diagnostic capabilities of these measurements. Full spectral resolution will be assumed in what follows, however. By considering a well-resolved Stokes spectrum, we are not losing generality; on the contrary, we are dealing with the most general problem of finding the possible diagnostics from individual wavelength samples. This aim will occupy us for most of this chapter. Consideration of those other measurements involving several wavelength samples can be made afterwards. In fact, we shall deal with that problem too in this chapter.

10.1 Probing the medium by scanning spectral lines

Light reaches the observer after having traveled through and interacted with the medium. Photons are emitted at a given point and then absorbed and re-emitted (or scattered) at another point in the medium. Since the distribution of energy (Stokes I) and the polarization state (Stokes Q/I, U/I, and V/I) are known to vary across the wavelength span of single spectral lines, it is conceivable that the different photons at the various wavelengths are "formed" in different parts of the medium, each with specific physical properties. If this conjecture is true (which, we assure the reader, qualitatively it is) one could probe the optical path by simply "tuning" into the different wavelength samples of the spectral line or by selecting several lines for analysis. In order to understand this issue let us begin by considering an isotropic medium, which will provide a perfect introduction of the concepts to be applied later to anisotropic media.

10.1.1 The isotropic case

Assume that we are dealing, for instance, with a non-magnetized stellar atmosphere. In this case, we already know that the solution of the RTE is

$$I(0) = \int_0^\infty \mathrm{e}^{-\int_0^{\tau_c} k(t)\,\mathrm{d}t} k(\tau_c)\, S(\tau_c)\, \mathrm{d}\tau_c, \qquad (10.1)$$

and that the evolution operator is the attenuation exponential (see Section 9.3). We

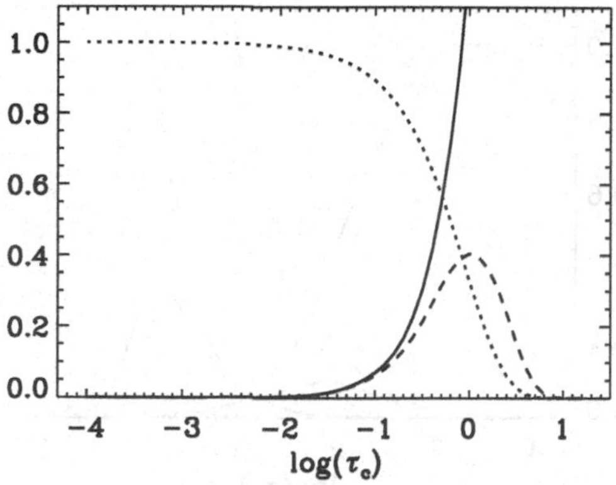

Fig. 10.1. Attenuation exponential (dotted line), emission coefficient (solid line), and contribution function (dashed line) of the Fe I line at 630.25 nm, at −11 pm from the line center. Both the emission coefficient and the contribution function have been multiplied by $\tau_c \ln 10$. Calculations have been made for a solar plage model without any magnetic field.

find fairly often in the astrophysical literature that $\log \tau_c$ is used as the integration variable instead of τ_c. This is done in order the better to sample (in numerical quadratures) the stratification of the atmosphere. Equation (10.1) can then be recast in the form

$$I(0) = \ln 10 \int_0^\infty \mathrm{e}^{-\ln 10 \int_0^{\tau_c} k(t)\, t\, \mathrm{d}(\log t)} k(\tau_c)\, S(\tau_c)\, \tau_c\, \mathrm{d}(\log \tau_c). \tag{10.2}$$

We keep τ_c as the integration variable in the following equations for the sake of simplicity. However, plots of the integrands will be made on the assumption that we are sampling the model atmosphere at equally spaced points in $\log \tau_c$. Hence, the relevant functions will be plotted multiplied by $\tau_c \ln 10$.

It is interesting to interpret the attenuation exponential as the probability of a photon, emitted at τ_c, for escaping freely (reaching zero optical depth) from the atmosphere. A glance to the dotted line in Fig. 10.1 helps in understanding this interpretation: the higher the photon is formed, the higher the escape probability. Photons formed very deep in the atmosphere (e.g., at $\log \tau_c = 1$) cannot reach the surface of the star, whereas those formed at, for example, $\log \tau_c = -2$ leave almost unhindered. The specific model atmosphere used to calculate the attenuation exponential in Fig. 10.1 is irrelevant since the shape of such a function in any other model is qualitatively the same.†

The product of the other two factors in the integrand of Eq. (10.1) directly gives the emission coefficient (solid line in Fig. 10.1), which is typically a monotonically

† Figures 10.1 to 10.5 of this section were originally conceived by B. Ruiz Cobo.

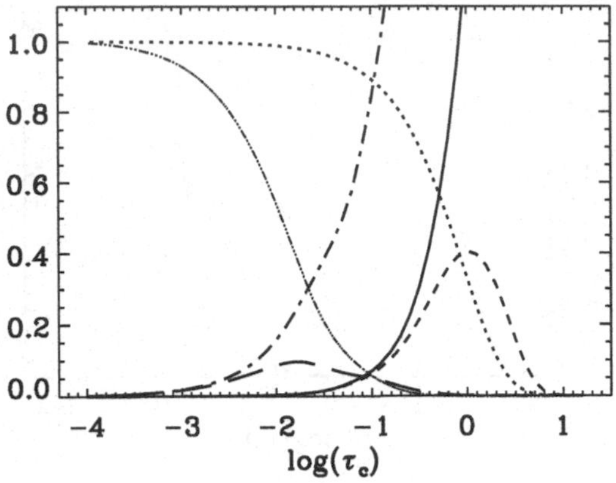

Fig. 10.2. Same as Fig. 10.1 for the wavelength samples at −11 nm and the very core of the line (dashed-triple-dotted, dashed-dotted, and long-dashed lines are correspondingly the same as dotted, solid, and short-dashed lines).

(exponentially) increasing function of $\log \tau_c$. Thus, the integrand of the formal solution of the RTE is an asymmetric bell-shaped function like the dashed line in Fig. 10.1 that results from multiplying the other two curves. When we run from the continuum wavelengths through the core wavelengths as in Fig. 10.2, we find that the bell-shaped function shifts towards higher layers of the atmosphere.

With this result in mind, we can interpret the integrand of the solution of the RTE (10.1) as a *contribution function*,

$$C(\tau_c) \equiv e^{-\int_0^\infty k(t)\,dt} k(\tau_c)\, S(\tau_c), \tag{10.3}$$

that informs us of how the different atmospheric layers *contribute* to the observed spectrum. With this interpretation, since the various contribution functions peak at different optical depths (see Fig. 10.3), one could say that *the closer to the line core, the higher the photons have mostly been formed.* This is a qualitatively true conclusion that holds for every absorption line whatever the model atmosphere. In fact, we can understand this result intuitively: the profile shape of the line indicates a variation of atmospheric opacity (or transparency) with λ. The medium is more transparent (light is less heavily absorbed) in the continuum wavelength region than in the line-core region. If the medium is more transparent at given wavelengths, then one can "see" deeper into the atmosphere at these wavelengths; in other words, the probability of photons for escaping from deeper layers is higher than at those wavelengths where the opacity is greater.

An extrapolation of the above results would induce us to think that some spectral lines are formed deeper and some other lines are formed higher in the atmosphere. This is a justified belief that, again, is qualitatively true. Weaker lines, that is, those

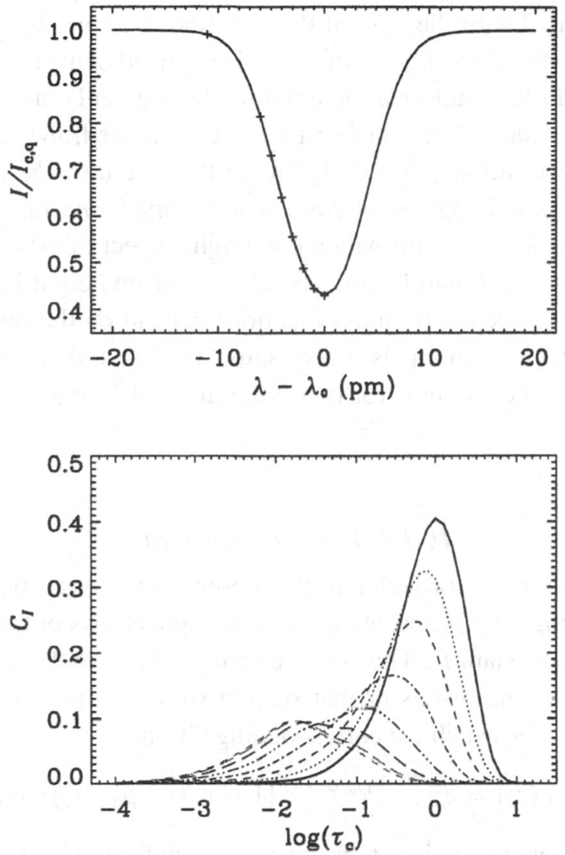

Fig. 10.3. Stokes I (top panel) sampled at several wavelengths (crosses) where the contribution function is evaluated (bottom panel). The contribution functions from right to left correspond to wavelength samples from left to right.

with a smaller depression, or those whose core is closer to the continuum intensity, are formed deeper on average than the stronger lines – those with significant depression cores. If one compares the core contribution functions of two lines, that with the stronger core has a contribution function (CF) that peaks at higher layers of the atmosphere. This type of comparison has driven numerous authors to believe that they are probing the various layers by simply measuring the same parameter over different spectral lines. Nevertheless, their reasoning fails when they attempt to extract quantitative information: our conclusions are qualitative and any quantitative extrapolation may be wrong. A spectral line whose minimum is 0.8 I_c has formed deeper on average than another line whose minimum is at 0.2 I_c, but by how much? *We cannot answer this question.* First, because the comparison may be strictly true for the two line cores, but certainly not for the wings of the second line when compared with the core of the first line. Second, because even within every one of the lines a broad range of layers is involved in the spectrum formation. Third, because CFs for single-wavelength samples are not Dirac delta distributions but have a finite width. This last fact means that, indeed, the single wavelength samples are formed in fairly broad regions of the medium: *the probability is not zero that two photons of a given wavelength may come from widely separated parts of the medium.* Hence, the information one might expect to extract from measurements at this wavelength hardly corresponds to a given height in the atmosphere. And finally because the contribution functions depend on the model atmosphere: one cannot say that a given line is "universally" formed at that height; its CFs may peak at different optical depths for two different model atmospheres, as we shall see later.

10.1.2 The anisotropic case

To better understand line formation in the presence of a magnetic field, let us proceed by considering light propagation along the optical axis of the medium, i.e., in the direction of a constant $\boldsymbol{B}$. This has been called (Section 9.4.2) the longitudinal case. According to the results of that section we have only to deal with Stokes $I+V$ and $I-V$, for which the corresponding CFs are:

$$C_{I\pm V}(\tau_c) = \mathrm{e}^{-\int_0^{\tau_c}(1+\eta_0\,\phi_{\mp1})\,\mathrm{d}t}[1+\eta_0(\tau_c)\,\phi_{\mp1}(\tau_c)]\,S(\tau_c). \tag{10.4}$$

Repeating similar calculations to those carried out for the isotropic case, but now with a longitudinal magnetic field whose strength varies linearly from 3600 G at $\log\tau_c = 1.2$ to 1000 G at $\log\tau_c = -4$, we find the results summarized in Fig. 10.4. At -10 pm from the line center, $I+V$ is made up of photons that have mostly formed deeper than those of $I-V$. Since integration is a linear operation and $V = 1/2(I+V) - 1/2(I-V)$, we find the CF for V to be

$$C_V(\tau_c) = \frac{1}{2}C_{I+V}(\tau_c) - \frac{1}{2}C_{I-V}(\tau_c), \tag{10.5}$$

which is shown in the bottom panel of Fig. 10.4. Note the significant difference between the contribution function for Stokes I and the contribution function for Stokes V. The latter has two lobes, one of them being negative! What does this mean? What kind of contributing photons provides a negative contribution? The appearance of (two or more) negative lobes in the CFs of Stokes Q, U, and V is natural since we know that all the three Stokes parameters have been defined as differences between two intensity measurements (see Chapter 3). What happens simply is that the interpretation of C_Q, C_U, and C_V as a "contribution" somehow loses its meaning: photons have been "added" at some place and are "subtracted"

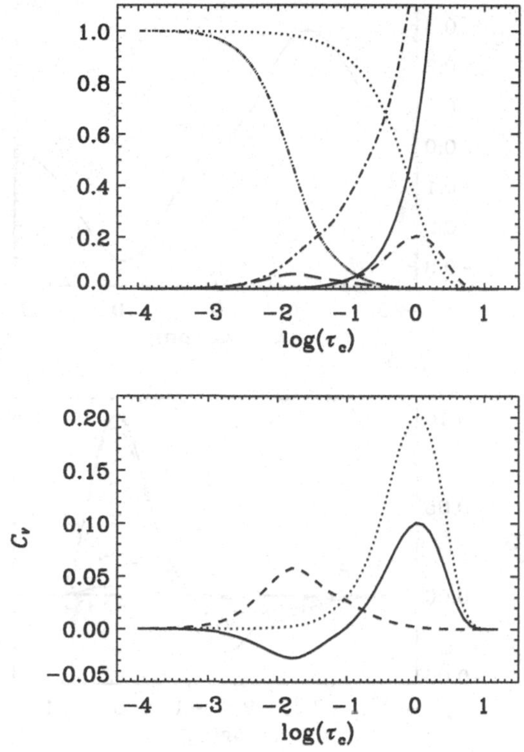

Fig. 10.4. Top panel: Attenuation exponential (dotted and dashed-triple-dotted lines), emission coefficient (solid and dashed-dotted lines), and contribution function (short- and long-dashed lines) of $I + V$ and $I - V$, respectively, of the Fe I line at 630.25 nm, at -10 pm from the line center. Calculations have been made in a solar plage model with a longitudinal magnetic field that varies linearly with $\log \tau_c$ from 3600 G at $\log \tau_c = 1.2$ to 1000 G at $\log \tau_c = -4$. Bottom panel: CFs of $I + V$ (dotted line), $I - V$ (dashed line), and V (solid line).

elsewhere; both the addition and the subtraction are important in reaching the specific state of polarization. Moreover, when we scan the spectral line, the modifications of C_V are not as clear as those for C_I (see Fig. 10.5). Quite remarkably, almost the whole photosphere seems to have an influence on the V profile. A mean depth of formation can hardly be ascribed to the various wavelength samples. Note that the sample at the central wavelength of the line has a zero contribution. This is a consequence of V being zero at that wavelength. Nevertheless, the reader should realize that every antisymmetric function (in this interval of $\log \tau_c$) could have been a CF for that wavelength as well.

In the general case, one obviously has

$$\boldsymbol{C}(\tau_c) = \mathbf{O}(0, \tau_c)\,\mathbf{K}(\tau_c)\,\boldsymbol{S}(\tau_c), \tag{10.6}$$

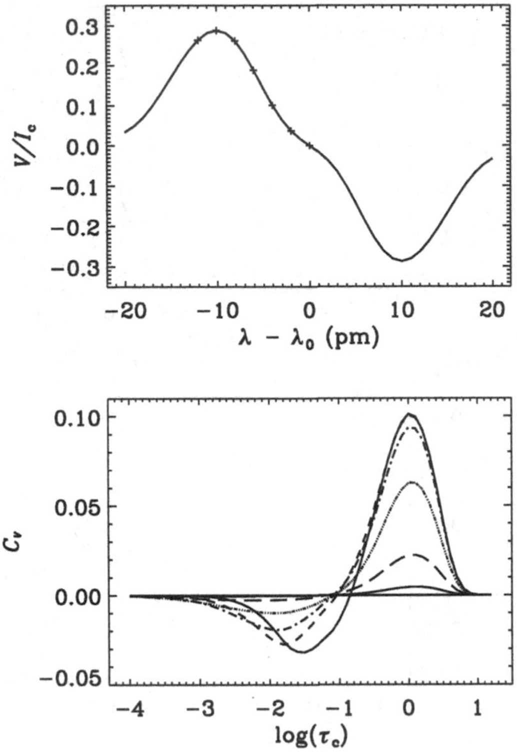

Fig. 10.5. Stokes V (top panel) sampled at several wavelengths (crosses) where the contribution function is evaluated (bottom panel). The contribution functions of the wavelength samples from left to right are represented by (in this order) solid (large), short-dashed, dashed-dotted, dashed-triple-dotted, long-dashed, solid (small), and straight, solid line equal to zero.

and the contribution function is a vector function with one component for each of the Stokes profiles. Equation (9.20) ensures that the integral of this contribution function indeed gives the emergent Stokes spectrum. A graphical illustration of such a general CF is shown in Fig. 10.6, where $\boldsymbol{C}$ is represented as a function of the logarithmic optical depth (on the X axis) and of the wavelength (on the Y axis). The calculations have been made for the Fe I line at 630.25 nm in the penumbral model of Fig. 9.3. $C_I(\lambda; \tau_c)$ is perhaps the most easily interpretable of the four components. Note that, far from the line center, the contribution function peaks at low atmospheric layers, whereas the central wavelengths show contributions from much higher layers. The Zeeman splitting is also visible, and, most importantly, the magnetic field strength gradient is easily discernible: the splitting is larger at lower atmospheric layers than at higher atmospheric layers. The LOS velocity gradient is more conspicuous in the other three components, where the wavelength asymmetry

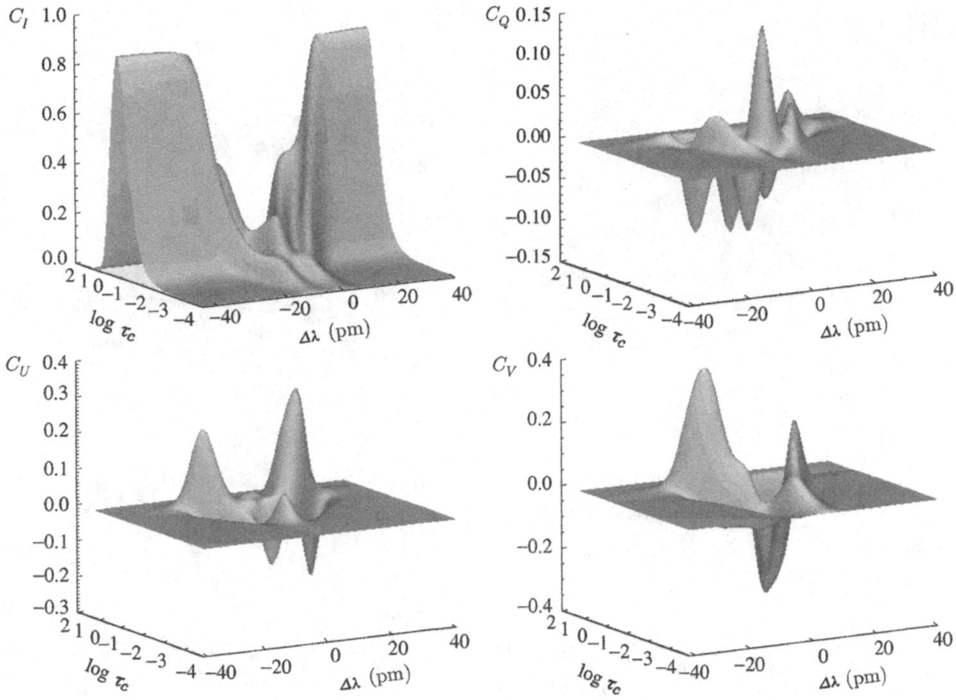

Fig. 10.6. Contribution function of the Fe I line at 630.25 nm in the penumbral model of Fig. 9.3. Wavelength is along the Y axis and logarithmic optical depth along the X axis.

of the CFs is noteworthy. We know from Section 9.2.1 that, in the absence of a velocity gradient, the solution of the RTE and, hence, the corresponding CFs have definite wavelength symmetry properties.

10.2 Height of formation of spectral lines and the fundamental ill-definition of the CFs

Many authors have tried to reach a sound definition of the height of formation for a given spectral line. Observers have long dreamed of the possibility of ascribing definite depths to the various lines ('this line is formed at x km above the reference and that other line is formed at y km...'). Were this ascription possible, gradients of physical quantities would be obtained by simply inferring the quantities by making use of the differently formed lines. Several attempts have been made but the most cursory glance at Figs 10.5 and 10.6 should discourage even the most optimistic researcher: a given spectral line receives "contributions" from a fairly broad range of layers, not from a single layer alone. This conclusion by itself is recommendation enough to refrain from looking for a height of formation of

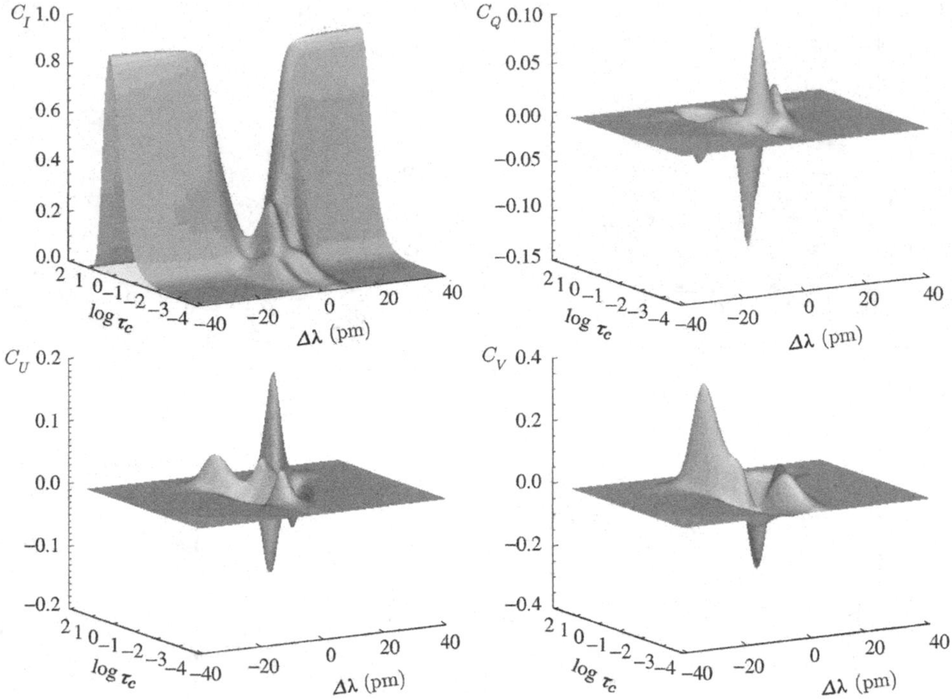

Fig. 10.7. Contribution function of the Fe I line at 630.25 nm in the penumbral model of Fig. 9.3 but with a constant magnetic field strength of 1000 G. Wavelength is along the Y axis and logarithmic optical depth along the X axis.

spectral lines. However, there are yet further reasons in the same spirit, as we shall see.

Very interestingly, CFs depend strongly on the model atmosphere. A given spectral line is formed differently in the quiet Sun and in a sunspot. Every physical quantity has an influence on the emergent Stokes spectrum and this influence varies from one atmosphere to another. As an example, consider the same Fe I line at 630.25 nm in Fig. 10.6 but now formed in a similar atmosphere with the sole difference of a constant magnetic field strength of 1000 G. The new contribution functions are plotted in Fig. 10.7. A comparison with Fig. 10.6 shows clear differences, although the two atmospheres are fairly close to each other.

Certainly, if we are interested in using the CFs to measure the magnetic field at given atmospheric heights, we have serious problems. In everyday language, we might say that "the ruler varies with the object to be measured" and is hence totally useless. The difficulties may be even greater when one is interested in finding a mean depth (or height) of formation for the equivalent width, the Stokes V

peak distance, or any other parameter involving several wavelengths across the line profile at the same time.

Furthermore, from a mathematical viewpoint, CFs are ill-defined functions. In fact, they are defined as integrands. Thus, any function with the same integral (i.e., whose integral is the Stokes spectrum) will be a CF in every respect. Specifically, every vector function of the form $\boldsymbol{C}+\boldsymbol{f}$, with $\boldsymbol{f}(\tau_c)$ being a function whose integral is zero, will be a contribution function since

$$\int_0^\infty [\boldsymbol{C}(\tau_c)+\boldsymbol{f}(\tau_c)]\,d\tau_c = \int_0^\infty \boldsymbol{C}(\tau_c)d\tau_c = \boldsymbol{I}(0). \tag{10.7}$$

In summary, CFs provide some interesting qualitative ideas on where and how each of the four Stokes parameters has been formed but they cannot be used to ascribe measurements to given atmospheric heights.

10.2.1 *Alternative CF definitions*

This basic failure of CFs to provide quantitative measurements of the height of formation of spectral lines was soon realized. It prompted theoreticians to propose alternative definitions.† Particularly noteworthy is the work of Magain (1986), who proposed to use a "line-depression contribution function" by formulating and solving (formally) a transfer equation for the line depression parameters $\boldsymbol{u}_0 - \boldsymbol{I}(0)/I_c(0)$, where $\boldsymbol{u}_0 \equiv (1,0,0,0)^{\mathrm{T}}$. It is very easy to see, however, that these new contribution functions, $\boldsymbol{C}_D$, are such that

$$\boldsymbol{u}_0 - \frac{\boldsymbol{I}(0)}{I_c(0)} \equiv \int_0^\infty \boldsymbol{C}_D(\tau_c)\,d\tau_c \tag{10.8}$$

and suffer from the same ill-definition as the formerly mentioned contribution functions, and that they are unable to provide any quantitative measure of the height of formation of spectral lines (see Ruiz Cobo and del Toro Iniesta, 1994). As a matter of fact, we shall see in the remainder of this chapter that the concept "height of formation of a spectral line" should not be trusted quantitatively: different physical quantities may be measured at different atmospheric heights with a given spectral line.

10.3 The sensitivities of Stokes profiles

In view of the ambiguities found in the search for a mean depth of formation of spectral lines, we had better change our minds and search in another direction. One is indeed interested in obtaining the diagnostic capabilities of the different

† The interested reader is directed to the paper by Ruiz Cobo and del Toro Iniesta (1994) for an extensive bibliography.

spectral lines. In other words, we can think of the spectral lines as the needle of an ammeter in the laboratory: when the intensity of the circuit changes, the needle moves; when the physical quantities of the medium change, the Stokes spectrum hopefully responds with detectable variations. A given spectral line will be a good diagnostic of a given physical quantity *if it forms differently in two atmospheres having different values of that physical quantity*, i.e., if the two values produce observable differences in the Stokes spectrum of the line. Thus, we are not interested for the moment in where the line forms but in whether it varies appreciably when some physical quantity changes. Note that this condition is necessary but not sufficient. A given line may be decidedly different when emerging from two atmospheres because of changes in several physical quantities and not in one alone: there may be *cross-talk* among the physical quantities of the medium. Therefore, the diagnostic problem turns out to be formidable and requires careful treatment. As anticipated in the introduction to this chapter, the only available tool is the radiative transfer equation for polarized light, which has the necessary links between light and the medium from which it is coming. Unfortunately, although the RTE is a linear differential equation, it involves non-linear functions (**K**, ***S***, and τ_c) of the model atmosphere: the calibration process becomes even more complicated.

10.3.1 Linearization of the RTE and response functions

Since non-linear problems are difficult to deal with, a first-order approximation often used in almost every field of theoretical and experimental physics is *linearization*. By linearization we understand a perturbative analysis in which *small* perturbations of the physical parameters of the model atmosphere will propagate "linearly" to *small* changes in the observed Stokes spectrum. To make the analysis quantitative, let us start by considering a model atmosphere like that of Eq. (9.1):

$$\boldsymbol{x}(\tau_c) = [x_1(\tau_c), x_2(\tau_c), \ldots, x_{m-1}(\tau_c), x_m(\tau_c)], \tag{10.9}$$

where $x_i(\tau_c)$ represent all the physical quantities characterizing the model (T, p_e, B, θ, etc.), that is, characterizing the propagation matrix, **K**, and the source function vector, ***S***. Let us assume as well the validity of the hypotheses summarized in Section 8.5.

Consider small perturbations, $\delta x_i(\tau_c)$, that induce small changes in **K** and ***S*** that, to a first order of approximation, can be written in the form

$$\delta \mathbf{K}(\tau_c) = \sum_{i=1}^{m} \frac{\partial \mathbf{K}}{\partial x_i} \delta x_i(\tau_c) \tag{10.10}$$

and

$$\delta \boldsymbol{S}(\tau_{\rm c}) = \sum_{i=1}^{m} \frac{\partial \boldsymbol{S}}{\partial x_i} \delta x_i(\tau_{\rm c}). \tag{10.11}$$

If these small changes in the propagation matrix and the source function vector lead to small changes, $\delta \boldsymbol{I}$, in the Stokes vector, then one can introduce all these modifications into the RTE to get

$$\frac{{\rm d}(\boldsymbol{I} + \delta \boldsymbol{I})}{{\rm d}\tau_{\rm c}} = (\mathbf{K} + \delta \mathbf{K})(\boldsymbol{I} + \delta \boldsymbol{I} - \boldsymbol{S} - \delta \boldsymbol{S}). \tag{10.12}$$

Since the perturbations are assumed to be small, only the first-order terms of Eq. (10.12) may be kept and, after taking the RTE into account, one obtains

$$\frac{{\rm d}(\delta \boldsymbol{I})}{{\rm d}\tau_{\rm c}} = \mathbf{K}(\delta \boldsymbol{I} - \delta \boldsymbol{S}) + \delta \mathbf{K}(\boldsymbol{I} - \boldsymbol{S}). \tag{10.13}$$

Now, if we define an *effective* source function vector

$$\tilde{\boldsymbol{S}} \equiv \delta \boldsymbol{S} - \mathbf{K}^{-1} \delta \mathbf{K}(\boldsymbol{I} - \boldsymbol{S}), \tag{10.14}$$

then we obtain a differential equation for the Stokes profile perturbations,

$$\frac{{\rm d}(\delta \boldsymbol{I})}{{\rm d}\tau_{\rm c}} = \mathbf{K}(\delta \boldsymbol{I} - \tilde{\boldsymbol{S}}), \tag{10.15}$$

that is formally identical to the RTE itself. Therefore, the solution to Eq. (10.15) must be formally the same as that for the RTE [Eq. (9.20)]:

$$\delta \boldsymbol{I}(0) = \int_0^\infty \mathbf{O}(0, \tau_{\rm c})\, \mathbf{K}(\tau_{\rm c}) \tilde{\boldsymbol{S}}(\tau_{\rm c})\, {\rm d}\tau_{\rm c}. \tag{10.16}$$

The only difference is the effective source function vector.

Note that we have taken the existence of the inverse, $\mathbf{K}^{-1}$, of the propagation matrix for granted. This is justified because reversibility of physical processes demands that **K** be non-singular. Absorption and dispersion, the two basic phenomena that constitute the propagation matrix, are non-depolarizing as we have already discussed in Section 7.4. Depolarizing phenomena appear only from emission. According to Landi Degl'Innocenti and Landi Deg'Innnocenti (1981), non-depolarizing processes do not show the typical irreversible character of depolarizing ones. Hence, **K** represents reversible phenomena and must be invertible.

Pursuing the analogy between the RTE and Eq. (10.15), a *contribution function* to the perturbations of the observed Stokes profiles can be defined as the integrand of Eq. (10.16):

$$\tilde{\boldsymbol{C}}(\tau_{\rm c}) \equiv \mathbf{O}(0, \tau_{\rm c})\, \mathbf{K}(\tau_{\rm c}) \tilde{\boldsymbol{S}}(\tau_{\rm c}). \tag{10.17}$$

Now, Eqs (10.14), (10.10), and (10.11) imply that $\tilde{\boldsymbol{C}}$ must decompose into a sum of terms such as

$$\tilde{\boldsymbol{C}}(\tau_c) = \sum_{i=1}^{m} \tilde{\boldsymbol{C}}_i(\tau_c) \equiv \sum_{i=1}^{m} \boldsymbol{R}_i(\tau_c)\, \delta x_i(\tau_c), \tag{10.18}$$

where, by definition, the response function vector, $\boldsymbol{R}_i(\tau_c)$, of the Stokes profiles to perturbations of the parameter x_i is:

$$\boldsymbol{R}_i(\tau_c) \equiv \mathbf{O}(0, \tau_c) \left[\mathbf{K}(\tau_c) \frac{\partial \boldsymbol{S}}{\partial x_i} - \frac{\partial \mathbf{K}}{\partial x_i} (\boldsymbol{I} - \boldsymbol{S}) \right]. \tag{10.19}$$

In terms of the response functions (RFs), the solution (10.16) can be recast in the form

$$\delta \boldsymbol{I}(0) = \sum_{i=1}^{m} \int_0^\infty \boldsymbol{R}_i(\tau_c)\, \delta x_i(\tau_c)\, d\tau_c. \tag{10.20}$$

Therefore, the final modification of the observed Stokes profiles is given by a sum of terms, each related to one physical quantity among those characteristic of the medium. Each term is an integral over the whole atmosphere of the perturbation of the physical parameter multiplied by the response function to that parameter. If x_k is modified by a unit perturbation in a restricted neighborhood of τ_0, then the values of $\boldsymbol{R}_k$ around τ_0 give us the ensuing variation of the Stokes spectrum. Since the observed Stokes parameters are usually measured relative to reference values (e.g., the continuum intensity of the – unpolarized – quiet Sun) the units of RFs are the direct inverse of their corresponding quantities. Hence, response functions to temperature perturbations will be expressed in K^{-1}, to magnetic field strength perturbations in G^{-1}, etc. Equation (10.20) suggests for RFs the role of partial derivatives of the Stokes spectrum with respect to the physical quantities of the model atmosphere. Such a role is even clearer when that equation is substituted (for numerical purposes) by a quadrature of coefficients c_j,

$$\delta \boldsymbol{I}(0) = \Delta(\log \tau_c) \ln 10 \sum_{i=1}^{m} \sum_{j=1}^{n} c_j \tau_j \boldsymbol{R}_i(\tau_j)\, \delta x_i(\tau_j), \tag{10.21}$$

where we assume that our atmosphere is sampled at n equally spaced points in the logarithm of the continuum optical depth. Then, if we include the coefficients $\Delta(\log \tau_c) \ln 10\, c_j \tau_j$ into the RFs (as one usually does in numerical and graphical representations), these appear as the coefficients of a *linear* expansion of $\delta \boldsymbol{I}(0)$ in terms of δx_i. Thus, within the linear approximation, *response functions directly give the sensitivities of the Stokes spectrum to perturbations of the physical conditions of the medium.* Graphical examples of RFs can be seen in Figs 10.8, 10.9, 10.10, and 10.11. They correspond to the sensitivities of the Fe I line at 630.25 nm

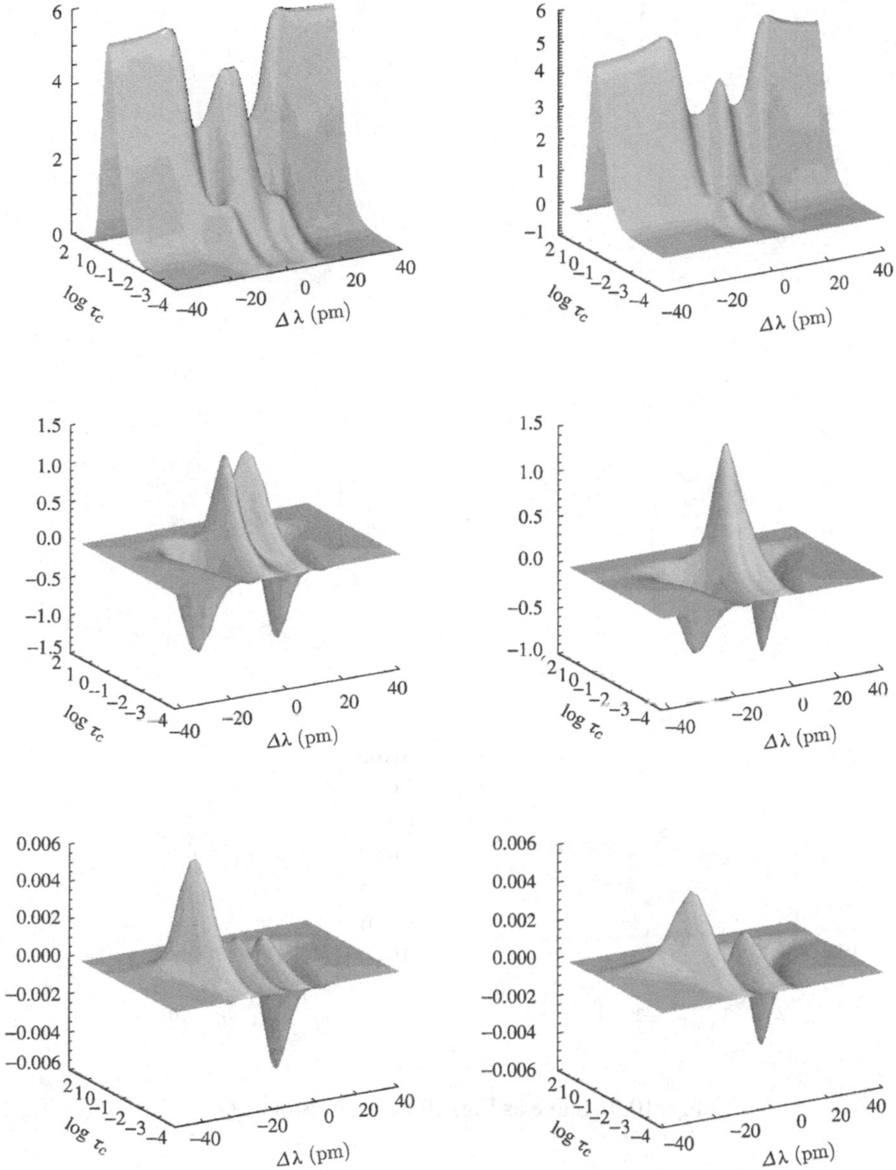

Fig. 10.8. Response functions of Stokes I of the Fe I line at 630.25 nm to perturbations of the temperature (top row), of the magnetic field strength (middle row), and of the line-of-sight velocity (bottom row) in two model atmospheres. The model of the left panels is 500 K hotter, has a magnetic field 500 G stronger, 20° more inclined, and with an azimuth 50° larger than the model of the right panels. The latter has a linear gradient of the LOS velocity whereas the former is at rest. Wavelength is along the Y axis (in pm with respect to the center of the line) and logarithmic optical depth along the X axis.

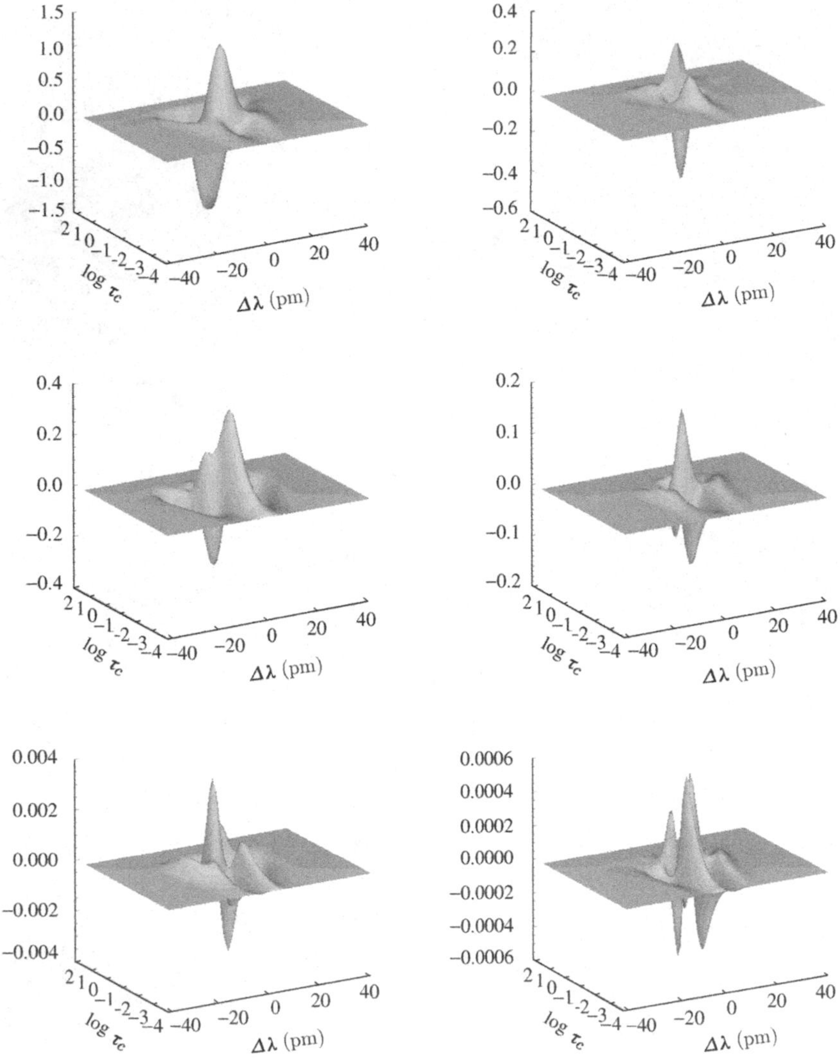

Fig. 10.9. Same as Fig. 10.8 but for Stokes Q.

in the HSRA model (Gingerich *et al.*, 1971) to which a constant $B = 2000\,\mathrm{G}$, $\theta = 30^\circ$, and $\varphi = 60^\circ$ has been added (left columns of the four figures) and in a model 500 K cooler, 500 G weaker, 20° less inclined, with an azimuth of 10°, and a linear stratification of the LOS velocity given by $1.58 + 0.3 \log \tau_c$ ($\mathrm{km\,s^{-1}}$). The Stokes profiles of this line in both model atmospheres are shown in Fig. 10.12.

A careful look at the RFs expression (10.19) easily reveals that the same evolution operator applies to the Stokes spectrum and to its linear perturbations. The terms in brackets represent those perturbations generated at τ_c that "evolve" up to

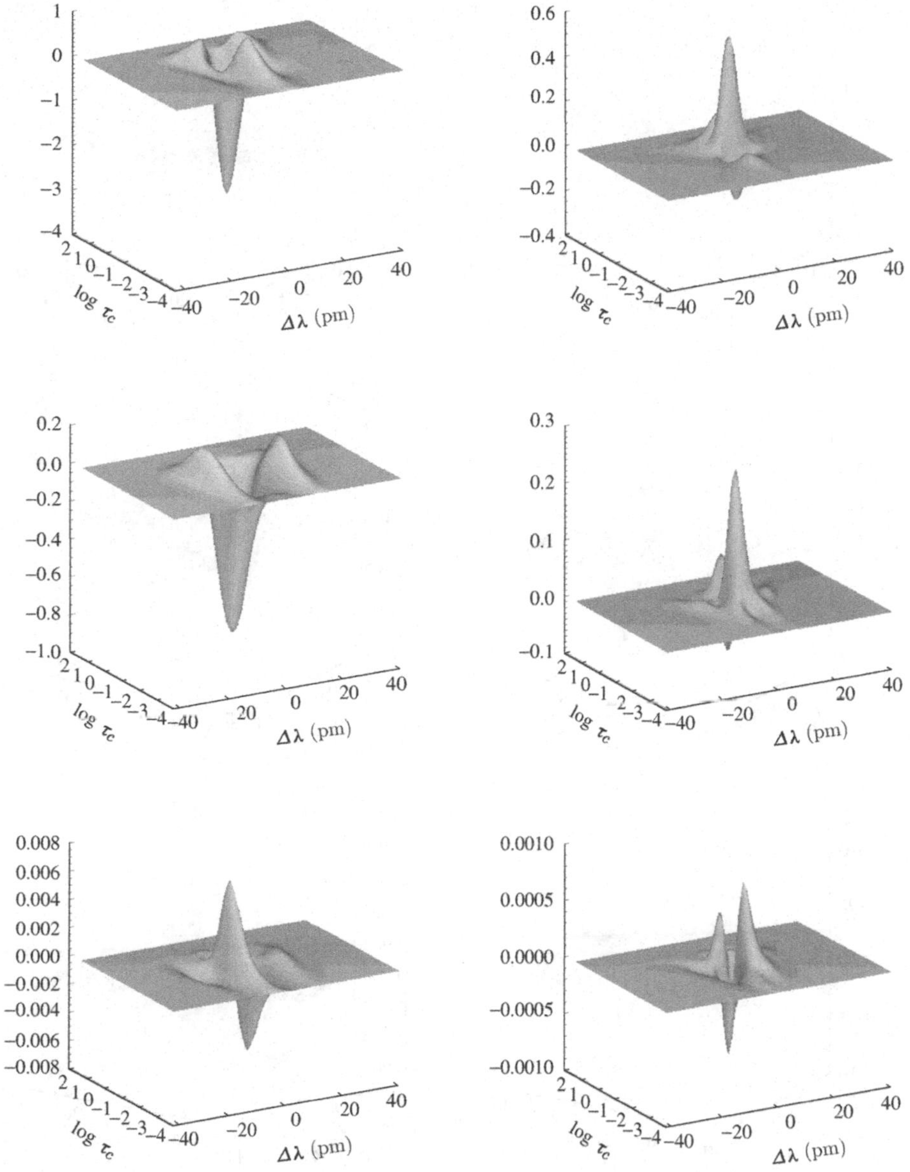

Fig. 10.10. Same as Fig. 10.8 but for Stokes U.

$\tau_c = 0$ by means of $\mathbf{O}(0, \tau_c)$. Notably, the perturbations follow both the variations of the propagation matrix and the variations of the source function vector. Obviously, line formation is influenced by the "sources" and the "sinks" of photons.†

† We already know that the propagation matrix not only accounts for the sinks of photons (absorption) but for dispersion effects as well.

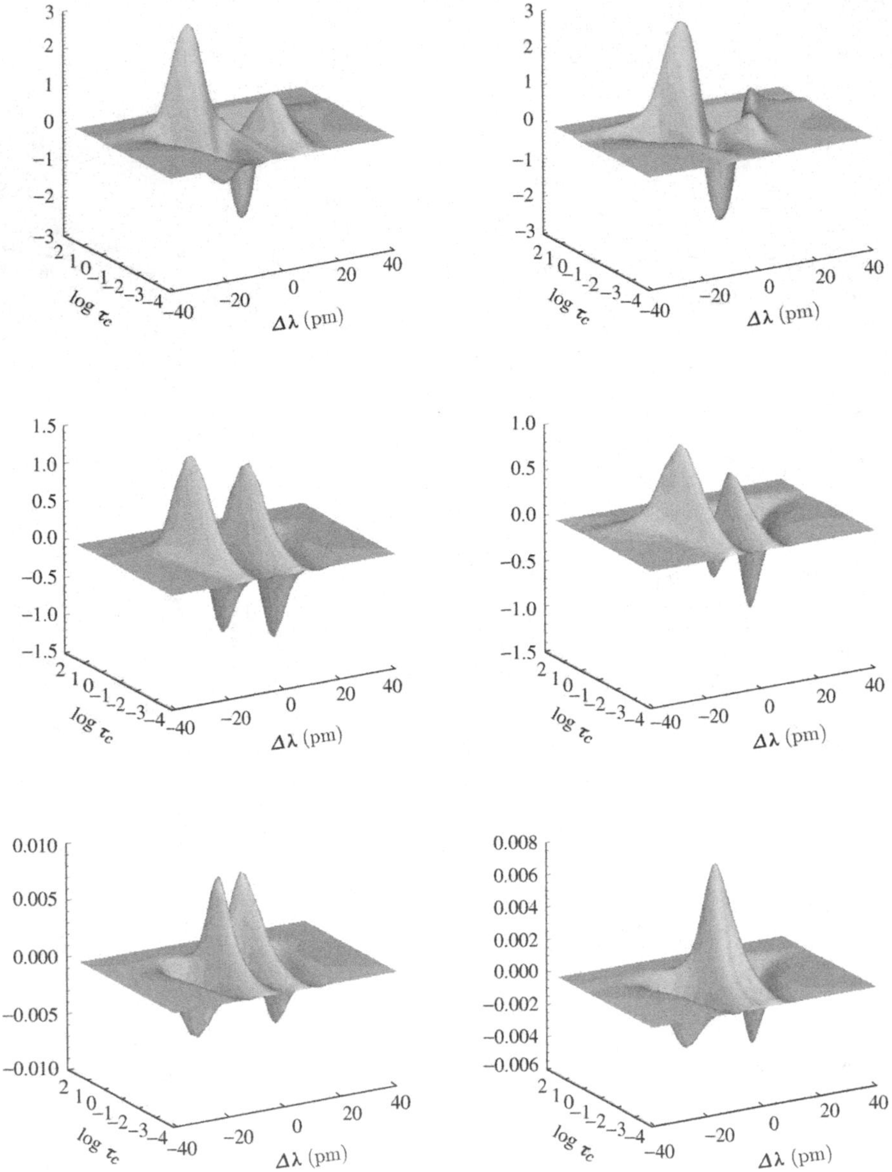

Fig. 10.11. Same as Fig. 10.8 but for Stokes V.

Moreover, the signs are changed for these two variations so that they compete against each other and may eventually cancel out. This is a very important fact that is sometimes forgotten in classical analyses which only account for **K** effects (absorption effects for unpolarized radiation). We return to this property later in this chapter.

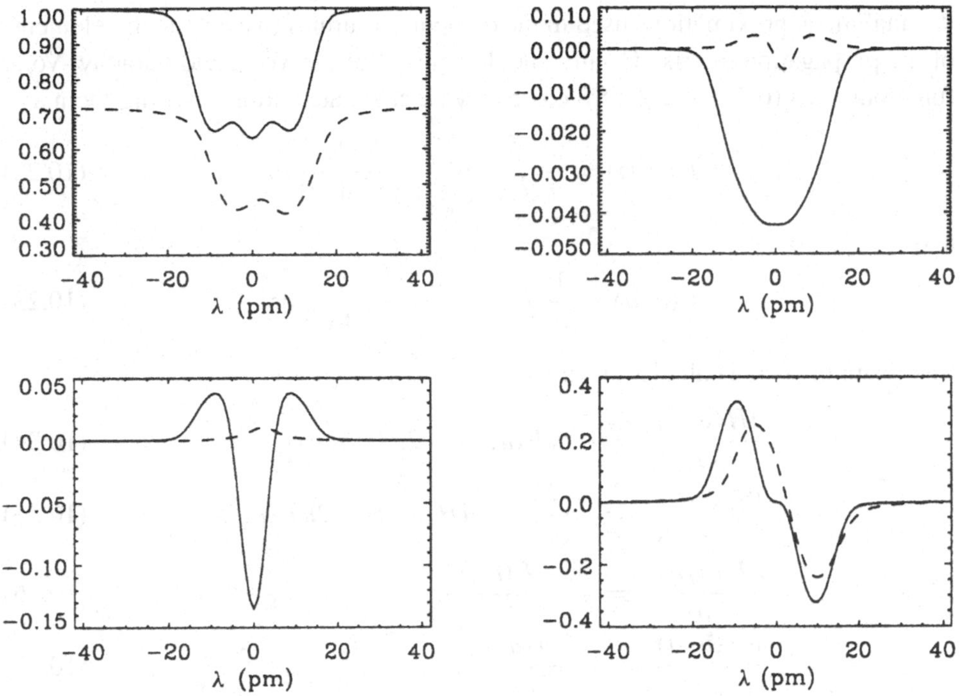

Fig. 10.12. Stokes profiles of the Fe I line at 630.25 nm as synthesized in the two model atmospheres of Figs 10.8, 10.9, 10.10, and 10.11. Solid lines correspond to the left-column model and dashed lines correspond to the right-column model.

Before continuing, the numerical evaluation of RFs deserves particular attention. RFs could certainly be calculated by a four-step process as follows: (1) synthesis of the Stokes spectrum in a given model atmosphere; (2) perturbation of *one* of the atmospheric parameters in a small neighborhood of optical depth and synthesis of a "perturbed" spectrum in the resulting model; (3) calculation of the ratio between the difference of both spectra and the perturbation of that parameter; (4) repetition of (2) and (3) for each optical depth, for each wavelength sample, and for the remaining atmospheric parameters. This is, of course, a long and tedious procedure that could be named "the brute force method". If, however, we realize that we have calculated (approximated) the evolution operator, the propagation matrix, the source function vector, and the Stokes profiles at each optical depth while synthesizing the observed spectrum, then it can easily be understood that we have all the ingredients to calculate the RFs. We only need the $\mathbf{K}$ and $\boldsymbol{S}$ derivatives. On the one hand, within the LTE approximation, the source function depends only on the local temperature. This dependence is well known analytically as the Planck function. On the other hand, we can calculate the derivative of the analytic approximations used in the computer routines to evaluate the elements of the propagation matrix. Finally, the derivatives of the Voigt and Faraday–Voigt functions [Eqs (6.57) and (6.58)] can easily be calculated after rewriting them as

$$H(u,a) = \frac{a}{\pi} \int_{-\infty}^{\infty} \mathrm{e}^{-(u-z)^2} \frac{1}{z^2+a^2}\, \mathrm{d}z \tag{10.22}$$

and

$$F(u,a) = \frac{1}{\pi} \int_{-\infty}^{\infty} \mathrm{e}^{-(u-z)^2} \frac{z}{z^2+a^2}\, \mathrm{d}z. \tag{10.23}$$

The result of such a calculation is:

$$\frac{\partial H(u,a)}{\partial u} = 2aF(u,a) - 2uH(u,a), \tag{10.24}$$

$$\frac{\partial F(u,a)}{\partial u} = \frac{2}{\sqrt{\pi}} - 2aH(u,a) - 2uF(u,a), \tag{10.25}$$

$$\frac{\partial H(u,a)}{\partial a} = -\frac{\partial F(u,a)}{\partial u}, \tag{10.26}$$

$$\frac{\partial F(u,a)}{\partial a} = \frac{\partial H(u,a)}{\partial u}. \tag{10.27}$$

The first two derivatives are directly calculated from Eqs (10.22) and (10.23) and after identifying the resulting terms. The fourth is obtained after Eq. (10.23) and integrating by parts. The third derivative is obtained after equating the second cross derivatives of F calculated from Eqs (10.25) and (10.27). Note that our expressions for these partial derivatives differ slightly from those by Ruiz Cobo and del Toro Iniesta (1994). The reason is that their definition of the Faraday–Voigt function is half our $F(u,a)$.

Equation (10.20) provides another interesting insight into the diagnostic problem: the modifications of the observed Stokes spectrum may be given by perturbations of different quantities or by perturbations of a single quantity but at various τ_c locations. That is, the same variation of $\boldsymbol{I}(0)$ may be produced by a change in temperature or by a change in the magnetic field strength, or even by a change in the magnetic field strength in the neighborhood of $\log \tau_c = 0$ or by a change in B in the region of $\log \tau_c = -3$. Hence, $\delta \boldsymbol{I}(0)$ cannot be ascribed to the perturbation of a given quantity at a given optical depth without considering all the remaining physical quantities characterizing the medium. Figures 10.8, 10.9, 10.10, and 10.11 illustrate very well the wealth of sensitivities that a single spectral line may have. The response to the various perturbations is very involved and the cross-talk among parameters is apparent. The techniques of analysis that try to disentangle information from the observed Stokes spectrum are really challenged by a fascinating and difficult diagnostic problem. Last, but not least, Eq. (10.20) readily explains that, unlike CFs, *response functions are not mathematically ambiguous*: the integrand of this equation is a product of two independent functions. In other words, whatever the procedure for evaluating RFs used, one should find the same result. As a matter of fact, this is the reason for using the "brute force method" for checking newly developed computer routines that calculate RFs directly through Eq. (10.19).

10.3.2 Properties of response functions

10.3.2.1 Relative sizes of relative RFs

A nice way of studying the relative importance of the various atmospheric parameters in line formation is to consider *relative* perturbations,

$$\delta x_i(\tau_c)/x_i(\tau_c).$$

Equation (10.20) may keep the same form if we rather use *relative* response functions, $\tilde{\boldsymbol{R}}_i(\tau_c) \equiv \boldsymbol{R}_i(\tau_c)\, x_i(\tau_c)$. Hence, $\tilde{\boldsymbol{R}}_i$ tells us about the reaction of the observed Stokes spectrum to relative (i.e., dimensionless) perturbations. Therefore, if one compares the various relative RFs, one easily discovers the relative importance of the atmospheric parameters in line formation. If, for instance, $\tilde{R}_{k,4}([\tau_0, \tau_1]) > \tilde{R}_{l,4}([\tau_0, \tau_1])$ at a given wavelength, we know that Stokes V at this wavelength depends more on the values of x_k than on the values of x_l over the region $[\tau_0, \tau_1]$ of optical depth. In general, experience shows that $\tilde{\boldsymbol{R}}_T$ is the largest RF at almost all depths and wavelengths: this is a natural consequence of T being the most important atmospheric parameter for line formation.

10.3.2.2 Sensitivities at various optical depths

As we probe the deeper atmospheric layers, the Stokes spectrum approaches the source function, since we have implicitly assumed that beneath the photosphere thermodynamic equilibrium prevails:

$$\lim_{\tau_c \longrightarrow \infty} (\boldsymbol{I} - \boldsymbol{S}) = \boldsymbol{0}. \tag{10.28}$$

Therefore, the second term in brackets of Eq. (10.19) vanishes at low optical layers. At these depths, the only relevant term is the source function derivative. Since we already know that under LTE, $\boldsymbol{S}$ depends only on T, we must conclude that RFs to temperature perturbations are significantly above zero down to much lower depths than any other RF, or that RFs to temperature have the slowest trend to zero at depth. In other words, the Stokes profiles may be sensitive to the temperature of lower layers even though they are completely insensitive to the values of the magnetic field vector or the LOS velocity at these lower layers.

Differences are expected not only in optical depth but also in wavelength when comparing the sensitivities to temperature, pressure, chemical composition, and atomic parameters with those to the magnetic field vector and the LOS velocity. This can be understood as follows. On the one hand, the propagation matrix can be recast as in Eq. (7.43),

$$\mathbf{K} = \mathbb{1} + \eta_0 \boldsymbol{\Phi},$$

where the elements of matrix $\boldsymbol{\Phi}$ can be identified from Eqs (8.14) and (8.15). On the other hand, η_0 does not depend on either $\boldsymbol{B}$ or $v_{\rm LOS}$. Thus, $\mathbf{K}$ derivatives with respect to B, θ, φ, and $v_{\rm LOS}$ will be of the form

$$\frac{\partial \mathbf{K}}{\partial x_i} = \eta_0 \frac{\partial \boldsymbol{\Phi}}{\partial x_i},$$

whilst $\mathbf{K}$ derivatives with respect to the other parameters will be of the form

$$\frac{\partial \mathbf{K}}{\partial x_i} = \frac{\partial \eta_0}{\partial x_i} \boldsymbol{\Phi} + \eta_0 \frac{\partial \boldsymbol{\Phi}}{\partial x_i}.$$

10.3.2.3 Resemblance of CFs and RFs to T

A glance at Figs 10.6 and 10.7 on the one hand, and Fig. 10.8 on the other, reveals similar forms of the contribution function and the response function to temperature perturbations. The resemblance does not occur by chance but has a clear physical reason: temperature is the "dominant" quantity of line formation so that variations in temperature are "felt" by photons wherever they form. We can understand this with a simple order-of-magnitude estimate of the ratio $\tilde{C}_1/C_1$ for Stokes I when we include only the temperature sensitivity in $\tilde{C}_1$.

Since $\mathbf{K}$ depends on temperature mostly through η_0, that ratio is of the order

$$\frac{\tilde{C}_1}{C_1} \simeq \left[\frac{\mathrm{d} \ln B_\nu}{\mathrm{d}T} - \frac{\partial \ln \eta_0}{\partial T} \frac{I - B_\nu}{B_\nu} \right] \delta T, \tag{10.29}$$

where we have already equated the first element of the source function vector with the Planck function. If we now assume the Wien approximation to the Planck function,

$$\frac{\mathrm{d} \ln B_\nu}{\mathrm{d}T} \simeq \frac{h\nu}{kT^2}, \tag{10.30}$$

and, according to Gray (1992),

$$\frac{\partial \ln \eta_0}{\partial T} \simeq \frac{h\nu}{kT^2} \tag{10.31}$$

as well for our neutral iron line. Expression (10.29) can then be rewritten as

$$\frac{R_{T,1}}{C_1} \simeq \frac{h\nu}{k} \frac{\delta T}{T^2} \left(1 - \frac{I - B_\nu}{B_\nu} \right). \tag{10.32}$$

Therefore, wherever the ratio $(I - B_\nu)/B_\nu$ is small, the ratio between the response function of Stokes I to temperature and the Stokes I contribution function should be inversely proportional to the square of temperature. This certainly happens in the wings of the lines where, according to the discussions in Section 10.2, we are seeing deeper, so that closer to the limit $I - B_\nu \longrightarrow 0$ (Section 10.3.2.2). As a corollary of this resemblance of CFs and RFs to temperature, one might say that any attempt to ascribe mean depths of formation to given spectral lines has a high probability of being wrong since CFs will mostly indicate the sensitivity to temperature. This last conclusion is yet more reason to abandon the search for mean depths of formation of spectral lines.

10.3.2.4 Wavelength symmetries

Applying similar reasonings to those in Section 9.2.1 to analyze the wavelength symmetries of the Stokes profiles leads to the conclusion that RFs keep the same symmetry properties: response functions of Stokes I, Q, and U are even functions of wavelength about the central wavelength of the line and response functions of Stokes V are odd functions of wavelength in the absence of LOS velocity gradients. Therefore, when material velocities are constant along the optical path, the same information is carried by both wings of the line. Wavelength samples that are symmetric with respect to the central wavelength of the line have the same sensitivities to perturbations of any atmospheric parameter. Information is redundant and the redundancy is removed only in the presence of velocity gradients.

10.3.2.5 Response functions to constant perturbations

There are physical parameters that can be assumed constant throughout the optical path. The perturbations to these parameters must be constant as well. In some instances, although the atmospheric parameter is not constant with depth, consideration of perturbations that are constant might be interesting. The RF to such constant perturbations of one of these parameters, for example x_k, is simply the integral over the whole atmosphere of $\boldsymbol{R}_k(\tau_c)$. Imagine that x_k is the only relevant quantity so that

$$\delta \boldsymbol{I}(0) = \int_0^\infty \boldsymbol{R}_k(\tau_c)\,\delta x_k(\tau_c)\,\mathrm{d}\tau_c. \tag{10.33}$$

Since $\delta x_k(\tau_c) = \delta x_k =$ constant,

$$\delta \boldsymbol{I}(0) = \delta x_k \int_0^\infty \boldsymbol{R}_k(\tau_c)\,\mathrm{d}\tau_c \equiv \delta x_k \boldsymbol{R}'_k. \tag{10.34}$$

The new parameter, $\boldsymbol{R}'_k$, can be said to be *the* RF of the Stokes spectrum to constant perturbations of the parameter x_k.

An interesting example of a constant parameter is the product Agf of the element abundance, the multiplicity of the lower level of the transition (also called the statistical weight of the line), and the oscillator strength of the line (see Section 6.3). Since $\boldsymbol{S}$ does not depend on Agf, we need only worry about the derivative of $\mathbf{K}$ with respect to this quantity.

Equations (6.62) and (6.63) tell us about the direct proportionality between the absorption and dispersion profiles and the product Nf of the number of absorbers or dispersers per unit volume and the oscillator strength of the line. We know that N is to be calculated through the Boltzmann and Saha equations, which give the populations within the LTE approximation. These equations establish a proportionality between N and Ag, so that both χ_α and $\tilde{\chi}_\alpha$, and hence η_0 [see Eq. (7.40)], are proportional to the product Agf. Now, since the propagation matrix can be written as in Eq. (7.43), the partial derivative of $\mathbf{K}$ with respect to Agf is

$$\frac{\partial K}{\partial Agf} = \frac{\partial \eta_0}{\partial Agf} \boldsymbol{\Phi} = \frac{\eta_0}{Agf} \boldsymbol{\Phi} = \frac{\mathbf{K} - \mathbb{1}}{Agf}, \tag{10.35}$$

whence

$$\boldsymbol{R}_{Agf} = -\mathbf{O} \frac{\mathbf{K} - \mathbb{1}}{Agf} (\boldsymbol{I} - \boldsymbol{S}). \tag{10.36}$$

Thus,

$$\delta \boldsymbol{I}(0) = \delta(Agf) \int_0^\infty \boldsymbol{R}_{Agf}(\tau_c)\, \mathrm{d}\tau_c \equiv \delta(Agf)\, \boldsymbol{R}'_{Agf}. \tag{10.37}$$

The last two equations open up the possibility of using the Sun as an atomic physics laboratory where atomic parameters such as the multiplicity of the lower level of the transition or the oscillator strength, and solar parameters such as elemental abundances, can be evaluated.

Another interesting example of an atmospheric parameter assumed to be constant with depth is the macroturbulence velocity, although it enters the Stokes spectrum synthesis in a different way from the other parameters. The sensitivity of the observed spectrum to perturbations of this parameter is also calculated in a slightly different way. According to Eq. (9.24), a change in the macroturbulence velocity modifies only the macroturbulence Gaussian function:

$$\delta G_{\mathrm{mac}} = \frac{\partial G_{\mathrm{mac}}}{\partial \xi_{\mathrm{mac}}} \delta \xi_{\mathrm{mac}}, \tag{10.38}$$

whence

$$\delta \boldsymbol{I}_{\mathrm{obs}} = \boldsymbol{I} * \delta G_{\mathrm{mac}} = \boldsymbol{I} * \frac{\partial G_{\mathrm{mac}}}{\partial \xi_{\mathrm{mac}}} \delta \xi_{\mathrm{mac}}. \tag{10.39}$$

The response function of the Stokes parameters to perturbations of the macroturbulence velocity can then be simply defined as

$$\boldsymbol{R}'_{\xi_{\rm mac}} \equiv \boldsymbol{I} * \frac{\partial G_{\rm mac}}{\partial \xi_{\rm mac}}. \tag{10.40}$$

10.3.2.6 Response to harmonic perturbations

Consider a harmonic perturbation of the parameter x_k:

$$\delta x_k(\tau_{\rm c}) = a\,{\rm e}^{2\pi {\rm i} s \tau_{\rm c}}, \tag{10.41}$$

where s stands for the spatial frequency and the amplitude is small enough so as to be represented to a first-order approximation. If no other atmospheric parameter is perturbed, the Stokes spectrum is modified by

$$\delta \boldsymbol{I}(0; s) = a \int_0^\infty \boldsymbol{R}_k(\tau_{\rm c})\,{\rm e}^{2\pi {\rm i} s \tau_{\rm c}} {\rm d}\tau_{\rm c}. \tag{10.42}$$

If we define $\boldsymbol{R}_k(\tau_{\rm c}) = \boldsymbol{0}, \forall\, \tau_{\rm c} < 0$, Eq. (10.42) can readily be rewritten as

$$\delta \boldsymbol{I}(0; s) = a\,\mathcal{F}[\boldsymbol{R}_k], \tag{10.43}$$

where, again, $\mathcal{F}$ represents the Fourier transform operator. Thus, if the perturbation is harmonic, the change of the Stokes spectrum is given by the Fourier transform of the corresponding response function.

10.3.2.7 Model dependence of RFs

We have so far not discussed one of the most evident properties of RFs that is indeed conspicuous in the graphical illustration of Figs 10.8, 10.9, 10.10, and 10.11. This omission does not mean that the remaining property is less important. On the contrary, *the model-dependence of RFs* is germane to our understanding of Stokes spectrum diagnostics. We cannot ascribe a mean height of formation with CFs. Likewise, we cannot quantitatively say with RFs that a given line is sensitive to one parameter and is insensitive to another. This cannot be done because RFs do depend on the model atmosphere. Some qualitative estimates can be made, but quantitative determinations can be carried out only once we know the medium and, remember, the astrophysical unknown is in fact the medium. RFs teach us many things and, most importantly, they can be used in quantitative analysis techniques (see Chapter 11), but cannot be simply used to find "heights of sensitivities" except for *a posteriori* cases where the model atmosphere is known.

10.4 A theoretical description of measurement

In the introduction to this chapter, we have already noted that many measurements are sometimes not just a resolved portion of the Stokes spectrum, but some given combination of parameters and/or wavelength samples. An overview of the diagnostic capabilities of the Stokes spectrum must certainly deal with the analysis of such general measurements. The best way to proceed in this endeavor is first to define properly what such a general measurement is. Once we have fixed the assumptions (which are implicit in most cases) we will understand the advantages and drawbacks of the different measured parameters. We shall show that the most general measurement over the Stokes spectrum, $\boldsymbol{I}(0; \lambda)$, can be described in terms of the action of two operators over $\boldsymbol{I}(0; \lambda)$.

Assume the observed Stokes spectrum, $\boldsymbol{I}(0; \lambda)$, to be a vector field that is indefinitely differentiable with continuity over the set of real numbers. This assumption is fairly common and indeed implicit in most measurements. The continuity of $\boldsymbol{I}(0; \lambda)$ and/or its derivatives is important for the sampling process to make sense. Most modern measurements start, in fact, by sampling and truncating the spectrum (only a limited wavelength range is observed).

Let $\boldsymbol{\Psi}$ be a linear and continuous functional whose image of the Stokes spectrum is a $4q$-element vector $\boldsymbol{y} = (y_1, y_2, \ldots, y_{4q})$:

$$\boldsymbol{y} \equiv \boldsymbol{\Psi}[\boldsymbol{I}(0; \lambda)]. \tag{10.44}$$

This operator is in charge of the sampling and truncation operations. These two operations are usually described by the multiplication of $I(0; \lambda)$, $Q(0; \lambda)$, $U(0; \lambda)$, and $V(0; \lambda)$ by a Dirac comb distribution and a rectangular function of width equal to the wavelength span of the measurements. Actually, real sampling also involves a convolution of the spectrum with a narrow rectangular function because the detector pixels have finite dimensions (we do not have infinite resolution). Since convolution is a linear and continuous operation, $\boldsymbol{\Psi}$ is still able to describe actual sampling. The image $\boldsymbol{y}$ may be, for example,

$$\begin{aligned} \boldsymbol{y} \;=\; & [I(\lambda_1), I(\lambda_2), \ldots, I(\lambda_{q-1}), I_{\rm c}, \\ & Q(\lambda_1), Q(\lambda_2), \ldots, Q(\lambda_{q-1}), Q_{\rm c}, \\ & U(\lambda_1), U(\lambda_2), \ldots, U(\lambda_{q-1}), U_{\rm c}, \\ & V(\lambda_1), V(\lambda_2), \ldots, V(\lambda_{q-1}), V_{\rm c}]^{\rm T}, \end{aligned} \tag{10.45}$$

where we have suppressed the $\tau_{\rm c} = 0$ indicator for notational simplicity, and where we understand that the $\boldsymbol{y}$ elements are indeed *actual* samples of the Stokes spectrum.

After sampling and truncating, some operations are carried out with the samples in order finally to get a given parameter, ζ. Let Ω be a differentiable operator that performs such operations:

$$\zeta \equiv \Omega[\boldsymbol{y}] = \Omega\{\boldsymbol{\Psi}[\boldsymbol{I}(0; \lambda)]\}. \tag{10.46}$$

As we discussed in Section 10.3, ζ will be useful if it varies when the atmospheric quantities are perturbed. The differentiability of Ω is then of paramount importance if we are interested in deducing the diagnostic capabilities of the measured parameter ζ. It is not a very restrictive condition because, in fact, it is verified by the vast majority of measurements usually made over the spectrum. Mathematically speaking, differentiability ensures the existence of the gradient

$$\nabla_y(\Omega) = \left(\frac{\partial \Omega}{\partial y_1}, \frac{\partial \Omega}{\partial y_2}, \ldots, \frac{\partial \Omega}{\partial y_{4q}} \right), \tag{10.47}$$

which links, to a first-order approximation, the modifications of the Stokes spectrum with changes in ζ:

$$\delta\zeta = \nabla_y(\Omega) \cdot \delta \boldsymbol{y}, \tag{10.48}$$

where the symbol $\cdot$ indicates the scalar product.

10.4.1 Generalized response functions

Now we only need to establish a connection between the perturbations, δx_i, and the modification, $\delta\zeta$, of our scalar parameter. This task is easy since the continuity and linearity of $\boldsymbol{\Psi}$ allow us to write

$$\delta \boldsymbol{y} = \boldsymbol{\Psi}[\delta \boldsymbol{I}(0; \lambda)] = \sum_{i=1}^{m} \int_0^\infty \boldsymbol{\Psi}[\boldsymbol{R}_i(\lambda; \tau_c)]\, \delta x_i(\tau_c)\, d\tau_c, \tag{10.49}$$

an equation which can be introduced into Eq. (10.48) to give

$$\delta\zeta = \sum_{i=1}^{m} \int_0^\infty \nabla_y(\Omega) \cdot \boldsymbol{\Psi}[\boldsymbol{R}_i(\lambda; \tau_c)]\, \delta x_i(\tau_c)\, d\tau_c. \tag{10.50}$$

Therefore, analogously to Eq. (10.20), we can clearly identify a *generalized* response function of ζ to perturbations of the physical parameter x_k:

$$R_k^\zeta(\tau_c) = \nabla_y(\Omega) \cdot \boldsymbol{\Psi}[\boldsymbol{R}_k(\lambda; \tau_c)]. \tag{10.51}$$

We are thus able to know (to first order) the sensitivities of ζ to the atmospheric parameters.

Had ζ been obtained through two linear operators (with Ω also linear), the expression of the RF simplifies to

$$R_k^\zeta(\tau_c) = \Omega\{\boldsymbol{\Psi}[\boldsymbol{R}_k(\lambda; \tau_c)]\} \tag{10.52}$$

since we are already within the linear approximation. An example of such a case is the integral of any Stokes profile over wavelength.

10.4.2 An example

To better understand the usefulness of the theoretical description we have made of measurements so far, let us consider a specific example.† Suppose that $\zeta = W$, the equivalent width of the line. In this case, $\boldsymbol{y} = [I(\lambda_1), I(\lambda_2), \ldots, I(\lambda_{q-1}), I_{\mathrm{c}}]$ and $W = \Omega[\boldsymbol{y}] = \Delta \sum_{i=1}^{q-1} [1 - I(\lambda_i)/I_{\mathrm{c}}]$, where Δ is the wavelength sampling interval. Hence,

$$\nabla_y(\Omega) = -\frac{\Delta}{I_{\mathrm{c}}^2} \left[I_{\mathrm{c}}, I_{\mathrm{c}}, \ldots, I_{\mathrm{c}}, -\sum_{i=1}^{q-1} I(\lambda_i) \right].$$

With these expressions, it is easy to find that the response functions of the equivalent width to perturbations of the atmospheric quantities are given by

$$R_k^W(\tau_{\mathrm{c}}) = \frac{\Delta}{I_{\mathrm{c}}^2} \sum_{i=1}^{q-1} \left[R_k^{\mathrm{c}}(\tau_{\mathrm{c}}) I(0; \lambda_i) - R_{k,1}(\tau_{\mathrm{c}}; \lambda_i) I_{\mathrm{c}} \right], \tag{10.53}$$

where R_k^{c} is the RF of the continuum intensity and, obviously, $R_{k,1}$ is the RF of Stokes I to perturbations of the parameter x_k.

Remarkably, the response functions of the equivalent width result from adding up $q - 1$ differences corresponding to the $q - 1$ samples of our observed I profile except for that of the continuum. In each term, the sensitivity of the continuum and that of the sample compete and may eventually cancel each other out. Since within the LTE approximation, the continuum intensity depends only on temperature, this competition is relevant only to the sensitivity to temperature, but equivalent widths have indeed been used traditionally as a diagnostic of the temperature.

Figure 10.13 shows three examples of generalized relative RFs of the equivalent width. The calculations have been done for the Fe I line at 630.25 nm in the model atmosphere of the right-column panels of Figs 10.8, 10.9, 10.10, and 10.11. The RF to temperature perturbations is plotted with a solid line, that to field strength perturbations with a dashed line, and that to LOS velocity perturbations with a dashed-dotted line. The two latter RFs are multiplied by a factor 30 in order to be comparable in size to the first RF: whereas we can read the Y axis of the plot as the modification δW produced by a 1% perturbation of T, it is the modification induced by a 30% perturbation of either B or v_{LOS}. This figure is an excellent illustration of several of the features that we anticipated theoretically in this and the previous sections.

- The relative sizes of the relative RFs readily tell us that W depends mostly on T. Temperature is not only the dominant quantity of line formation but the main parameter for the equivalent width of the lines.

† The interested reader may find this and other examples of generalized RFs in the paper by Ruiz Cobo and del Toro Iniesta (1994).

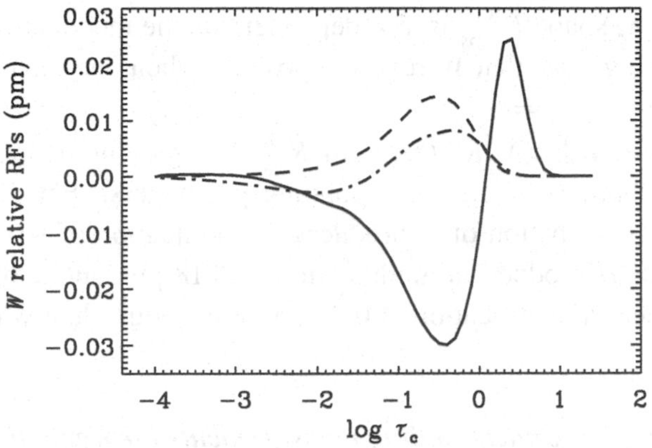

Fig. 10.13. Generalized response functions of the equivalent width of the Fe I line at 630.25 nm to relative perturbations (in percent) of the temperature (solid line), the magnetic field strength (dashed line), and the LOS velocity (dashed-dotted line). The two latter are multiplied by 30. The calculations have been carried out with the model atmosphere corresponding to the right-column panels of Figs 10.8, 10.9, 10.10, and 10.11.

- The double lobe of $\tilde{R}_T^W$ is a clear proof of both the greater sensitivity of the line to T at the very low layers of the atmosphere and of the competition between $\mathbf{K}$ and $\boldsymbol{S}$ in the final modification of W. Below $\log \tau_c = 0$, the only significant sensitivity of W is that to T. We already know (Section 10.3.2.2) that, at these low layers, only the $\boldsymbol{S}$ derivative with respect to temperature contributes to the response function; no variation in $\mathbf{K}$ can be felt by the spectrum. The positive lobe corresponds to the effects of an increase in the photon supply: if T is increased, the number of available photons to be absorbed in the layers above augments because the source function certainly increases. In contrast, the negative lobe shows a dominant role in the high layers of the $\mathbf{K}$ derivative, which implies a line weakening (a decrease in W) after an increase in temperature. This weakening of Fe I lines in the solar atmosphere† due to an increase of T was known from classical analyses which only took the absorption properties (for unpolarized light) into account. However, these analyses are unable to explain, for instance, the center-to-limb variation of spectral lines as was explained for the first time by Ruiz Cobo and del Toro Iniesta (1994).
- The line is insensitive to the values of both B and $v_{\rm LOS}$ at layers below $\log \tau_c = 1$.
- $\tilde{R}_B^W$ shows just a single positive lobe in the model atmosphere of our example. This means that W can be enhanced through an increase in B: the magnetic field can desaturate the line. The sensitivity of this very line to B perturbations may be different in other model atmospheres.

† Metals are mostly ionized in the solar photosphere. The most abundant ionization state of iron is Fe II.

- The double lobe shape $\tilde{R}^W_{v_{\rm LOS}}$ is also dependent on the model atmosphere. In this particular case, we see that W reacts oppositely when one increases $v_{\rm LOS}$ either above or below $\log \tau_{\rm c} = -1$.
- The integrals of $\tilde{R}^W_T(\tau_{\rm c})$, $\tilde{R}^W_B(\tau_{\rm c})$, and $\tilde{R}^W_{v_{\rm LOS}}(\tau_{\rm c})$ give the response to constant relative perturbations of the corresponding parameters. For this model atmosphere, a 1% perturbation of T produces a modification of −0.247 pm, a 30% perturbation of B produces a modification of 0.18 pm, and a 30% perturbation of $v_{\rm LOS}$ produces a modification of 0.06 pm of the equivalent width.

10.4.3 Understanding measurements theoretically

Many measurements like those described in Section 10.4 are believed to represent some kind of average of a given atmospheric parameter over the whole atmosphere. For example, the Stokes V peak distance is assumed to be proportional to the magnetic field strength within some interval of values; the magnetograph signal, that is, the integral of Stokes V over a narrow wavelength interval, is assumed to be proportional to the longitudinal component, $B \cos\theta$, of the magnetic field within some other interval of B values; the position of the minimum of Stokes I is often used as a measure of the LOS velocity; etc. All these parameters are single-valued whereas the corresponding physical quantities may vary throughout the optical path. We are then led to assume that the stratification of $B(\tau_{\rm c})$, $v_{\rm LOS}(\tau_{\rm c})$, etc., is somehow weighted and averaged to finally produce the measured parameter that hopefully corresponds to the actual value of the physical parameter at some given height. Certainly, the result of the average depends on the atmospheric parameter (B may have a completely different stratification than $v_{\rm LOS}$) and on the measurement itself (the technique of measuring Stokes V peak distances is different from that of integrating the profile).

Let us quantify all this reasoning and try to find a theoretical calibration (we are interested in obtaining the average that is usually assumed but unknown). This knowledge of the average will help in understanding the results of the measurements but, in any case, we do not pretend to find *a priori* recipes: the dependence of any measurement on the actual stratification of the atmosphere implies that only *a posteriori* results are expected.

Assume, for instance, that the measured parameter ζ, defined in Eq. (10.46), gives an estimate of the atmospheric parameter $x_k(\tau_{\rm c})$. We understand in this way that the necessary calibration factor for measuring ζ in the same units as x_k is already included in the specific definition of Ω. Assume further that the measurement is unbiased for constant stratifications of x_k, that is,

$$\zeta = x_{k,c} \tag{10.54}$$

whenever

$$x_k(\tau_c) = x_{k,c} = \text{constant}. \tag{10.55}$$

If we accept that ζ is representative of some average of the actual stratification, we are implicitly somehow looking for an "equivalent model atmosphere" where $x_k(\tau_c) = x_0$ is constant. Let ζ_0 be the result of applying the two operators $\boldsymbol{\Psi}$ and Ω to the profiles emerging from that "equivalent atmosphere". Then, the difference $\delta\zeta = \zeta - \zeta_0$ is given by

$$\zeta - \zeta_0 = \int_0^\infty R_k^\zeta(\tau_c)\,\delta x_k(\tau_c)\,\mathrm{d}\tau_c, \tag{10.56}$$

according to Eqs (10.50) and (10.51), where δx_k is the difference between the actual stratification and x_0:

$$\delta x_k(\tau_c) = x_k(\tau_c) - x_0. \tag{10.57}$$

The condition expressed in Eqs (10.54) and (10.55) necessarily implies that

$$\zeta_0 = x_0 \tag{10.58}$$

and

$$\int_0^\infty R_k^\zeta(\tau_c)\,\mathrm{d}\tau_c = 1, \tag{10.59}$$

that is, the generalized response function of ζ to perturbations of x_k must be normalized in area. Equation (10.58) is directly the hypothesis (10.54). Equation (10.59) comes from the fact that when $x_k(\tau_c) = x_{k,c} = \text{constant}$,

$$\zeta - \zeta_0 = (x_{k,c} - x_0)\int_0^\infty R_k^\zeta(\tau_c)\,\mathrm{d}\tau_c, \tag{10.60}$$

and the unbiasing condition requires

$$\delta\zeta = x_{k,c} - x_0,$$

whence Eq. (10.59).

The normalization of the generalized response function allows us to rewrite Eq. (10.56) in the form

$$\zeta = \int_0^\infty R_k^\zeta(\tau_c)\,x_k(\tau_c)\,\mathrm{d}\tau_c, \tag{10.61}$$

which gives us the measured parameter as an *average of the actual stratification weighted by the corresponding generalized response function*. Therefore, we have found what we were looking for. Note, however, that we have been implicitly assuming that the only atmospheric quantity having an influence on ζ is $x_k(\tau_c)$. This is hardly the case in practice and some *cross-talk* among the atmospheric parameters may appear.

10.4.3.1 *Heights of formation for measurements*

As we had foreseen, the result of the measurement depends on both the actual atmosphere (through R_k^ζ and x_k) and on the technique (through R_k^ζ). Note that only when we know $x_k(\tau_c)$ can we explore further the meaning of our measurement. However, if we have a good enough approximation to the actual atmosphere, we may even get an idea of a *height of formation* for that measurement. R_k^ζ does not depend strongly on the specific atmosphere: the response function is fairly similar in two atmospheres that are not too different from each other. Thus, we can assume a given stratification for $x_k(\tau_c)$ and find that optical depth τ_f at which the measurement and the stratification coincide:

$$\zeta \equiv x_k(\tau_f). \tag{10.62}$$

Equation (10.62) is the definition of the height of formation for ζ. If $x_k(\tau_c)$ is constant, τ_f has no meaning, of course. Assume that x_k is a continuous monotonic function of the optical depth. It then has an inverse function, x_k^{-1}, whose application to ζ gives τ_f:

$$\tau_f = x_k^{-1}(\zeta). \tag{10.63}$$

Therefore, Eq. (10.61) implies that

$$\tau_f = x_k^{-1}\left\{\int_0^\infty R_k^\zeta(\tau_c)\, x_k(\tau_c)\, \mathrm{d}\tau_c\right\}. \tag{10.64}$$

If, for instance, we assume that x_k depends linearly on τ_c, then the height of formation of measured parameter ζ turns out to be the barycenter of the generalized response function.

An illustration of the use in practice of these heights of formation for measurements is shown in Fig. 10.14, taken from a paper by Westendorp Plaza *et al.* (1998). It shows the stratification with the logarithm of the optical depth of the line-to-continuum absorption coefficient ratio, η_0, the Doppler width of the line, $\Delta\lambda_D$, and the damping parameter, a, in the HSRA model with $B = 1000\,\mathrm{G}$, $\theta = 60°$, $\varphi = 80°$, and a zero macroturbulence (left panels) or $\xi_{\mathrm{mac}} = 0.6\,\mathrm{km\,s^{-1}}$ (right panels). The authors carried out a numerical experiment where synthetic profiles (in these model atmospheres) fed the Milne–Eddington inversion technique that, among other things, provides measurements of these parameters. The dashed arrows indicate the theoretical heights of formation for the measurements whereas the solid arrows point to the optical depths where the values actually retrieved by the technique coincide with the stratification of the quantity. Note the very nice agreement between theoretical predictions and experimental results, although some cross-talk between the parameters induce measurable deviations. The larger deviations in the right-column panels arise from the presence of macroturbulence that is not foreseen by the Milne–Eddington technique.

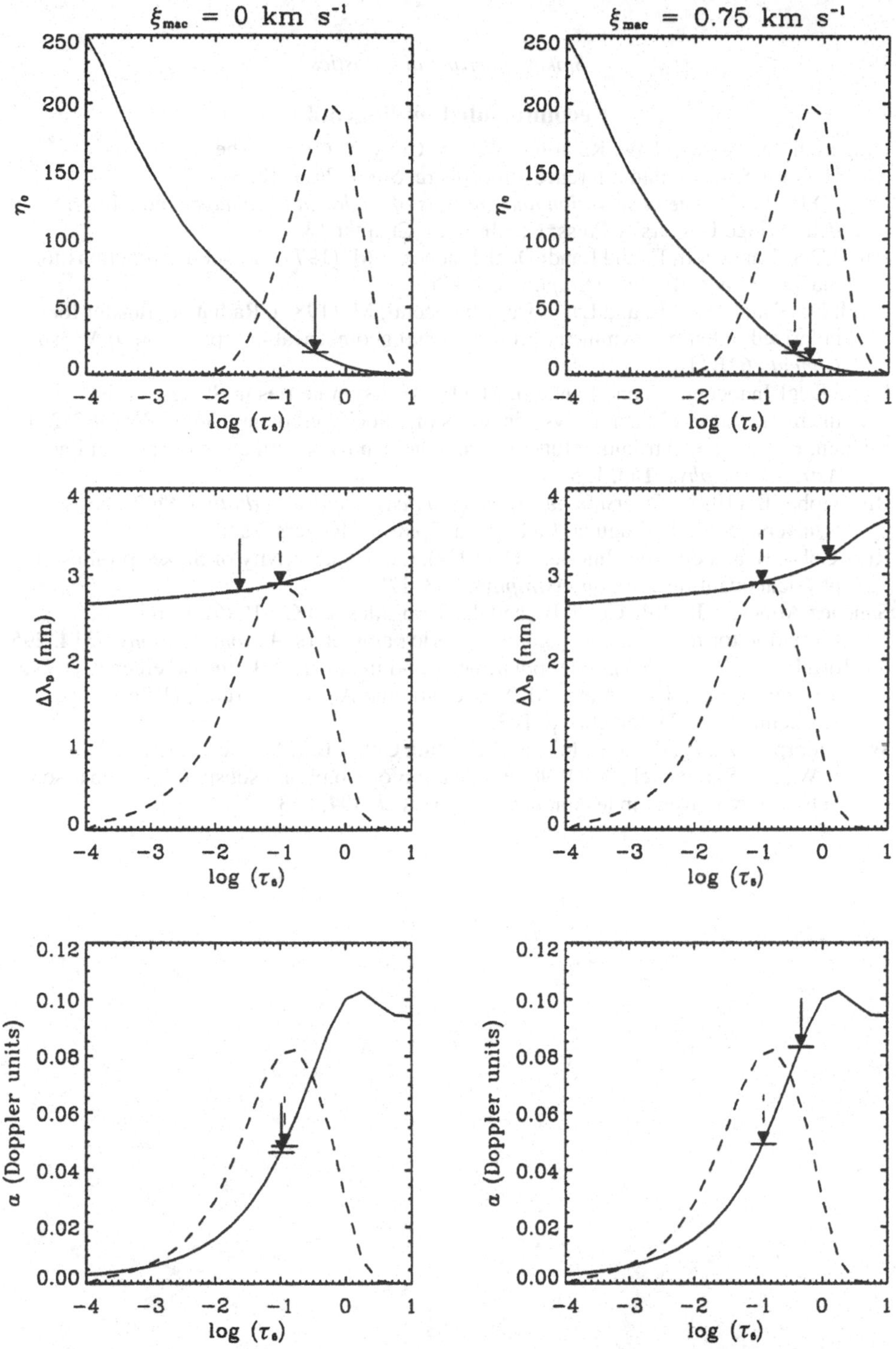

Fig. 10.14. Stratification of η_0, $\Delta\lambda_D$, and a (solid lines) and generalized response functions (dashed lines) of the Milne–Eddington inversion technique to perturbations of these quantities. Dashed arrows point to the predicted heights of formation, and solid arrows to the retrieved values from the Milne–Eddington technique. The two columns differ only in the macroturbulence velocity. The results come from numerical experiments. From Westendorp Plaza *et al.* (1998).

Recommended bibliography

Gingerich, O., Noyes, R.W., Kalkofen, W., and Cuny, Y. (1971). The Harvard-Smithsonian reference atmosphere, *Solar Phys.* **18**, 347.

Gray, D.F. (1992). *The observation and analysis of stellar atmospheres*, 2nd edition (Cambridge University Press: Cambridge). Chapter 13.

Landi Degl'Innocenti, E. and Landi Degl'Innocenti, M. (1977). Response functions for magnetic lines. *Astron. Astrophys.* **56**, 111.

Landi Degl'Innocenti, E. and Landi Degl'Innocenti, M. (1981). Radiative transfer for polarized radiation: symmetry properties and geometrical interpretation, *Il Nuovo Cimento* **62B**, 1.

Landi Degl'Innocenti, E. and Landolfi, M. (1983). Asymmetries in Stokes profiles of magnetic lines: a linear analysis in terms of velocity gradients. *Solar Phys.* **87**, 221.

Magain, P. (1986). Contribution functions and the depths of formation of spectral lines. *Astron. Astrophys.* **163**, 135.

Ruiz Cobo, B. (1992). *Inversión de la ecuación de transporte radiativo*. PhD Thesis (Universidad de La Laguna: La Laguna, Spain). Chapters 2 and 3.

Ruiz Cobo, B. and del Toro Iniesta, J.C. (1994). On the sensitivity of Stokes profiles to physical quantities. *Astron. Astrophys.* **283**, 129.

Sánchez Almeida, J., Ruiz Cobo, B., and del Toro Iniesta, J.C. (1996). Heights of formation for measurements of atmospheric parameters. *Astron. Astrophys.* **314**, 295.

del Toro Iniesta, J.C. (2001). Solar polarimetry and magnetic field measurements, in *The dynamic Sun*. A. Hanslmeier, M. Messerotti, and A. Veronig (eds.) (Kluwer Academic Pub.: Dordrecht), p. 183.

Westendorp Plaza, C., del Toro Iniesta, J.C., Ruiz Cobo, B., Martínez Pillet, V., Lites, B.W., and Skumanich, A. (1998). Optical tomography of a sunspot. I. Comparison between two inversion techniques. *Astrophys. J.* **494**, 453.

11
Inversion of the RTE

—"Would you tell me, please, which way I ought to go from here?", said Alice. —"That depends a good deal on where you want to get to", said the Cat. —"I don't much care where", said Alice. —"Then it doesn't matter which way you go", said the Cat.

—*Lewis Carroll, 1865.*

Once one has a solution of the RTE, the most simple procedure of inference can be devised such that a comparison between calculated and observed Stokes spectra suggests modifications in prescribed models of the medium. Iteratively improved models refine the match between observations and theoretical calculations. When the match is good enough, the last model in the iteration is taken as a model of the medium and its characteristic parameters are the inferred parameters of the medium. This trial-and-error method may be useful when the model medium is very simple and contains just a few free parameters. Note that every change of a given free parameter implies an integration of the RTE which is a process requiring a great deal of computer time. If the number of free parameters is large, the manual trial-and-error method can become impracticable, but even automated trial-and-error procedures that modify the various parameters randomly (blindly) may not converge to a physically reasonable final model of the medium. The results may even seem reasonable but be greatly in error. As an illustration of this last possibility, Fig. 11.1 shows the Stokes V profile of the Fe I line at 630.25 nm synthesized in two model atmospheres. The open circles correspond to a model 2000 K cooler than the penumbral model of del Toro Iniesta *et al.* (1994) already used in previous chapters. To this model in hydrostatic equilibrium, a constant longitudinal magnetic field of 800 G and a macroturbulence velocity of 1 km s^{-1} have been added. The solid line corresponds to a similar model but 300 K hotter, with a magnetic field weaker by 270 G, and with a macroturbulence velocity 0.185 km s^{-1} higher. If the first profile were the observations and the second profile the trial-and-error

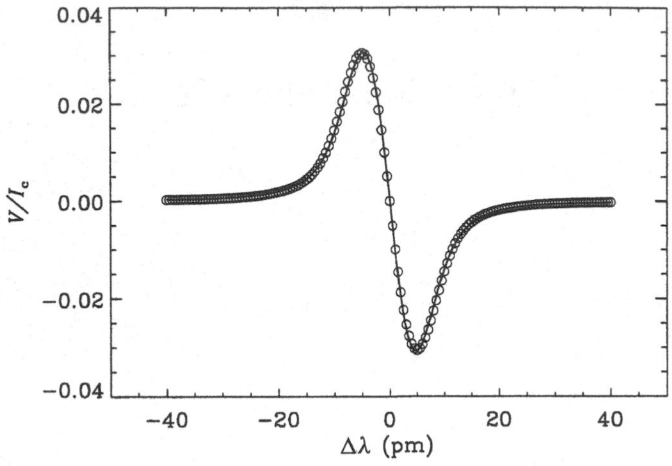

Fig. 11.1. Open circles: Stokes V profile of the Fe I line at 630.25 nm synthesized in a model atmosphere in hydrostatic equilibrium, 2000 K cooler than the del Toro Iniesta *et al.* (1994) model, with a constant longitudinal magnetic field of 800 G and a macroturbulence velocity of 1 $\mathrm{km\,s^{-1}}$. Solid line: Stokes V profile of the same line, synthesized in a model atmosphere 300 K hotter that the former, 270 G weaker, and with a higher macroturbulence velocity of 1.185 $\mathrm{km\,s^{-1}}$.

fit, the inferred model atmosphere should be completely wrong. We can attribute the wrongness of the result to *cross-talk* among the free parameters: in this particular case, an increase in T produces similar effects to a decrease in B plus a touch of ξ_{mac}. This cross-talk is not always the same but, in general, one must be cautious when trying to reproduce the observations by sequentially modifying the various free parameters.

An extreme example of a very simple model and inference is provided by the measurement of material velocities through that of line-core shifts. Here, the model is such that no other physical quantity is thought to have an influence on the position of the line, and the line-of-sight velocity is assumed constant throughout the optical path. The model is so simple that no iterative procedure is even needed: the core shifts can be directly calibrated to velocity units by means of the Doppler formula. No further considerations are needed. However, simplicity does not imply suitability for every particular observation: the hypotheses may be inapplicable to given observed features. For example, the meaning of the measurement is not so clear when a gradient of velocity with depth is present in the atmosphere. On the other hand, we may be able to measure velocities but may also be interested in magnetic fields, temperatures, etc. Therefore, more elaborate models are often used because they cover broader ranges of applicability and allow us to retrieve several free parameters at the same time. An example is the reproduction of the four Stokes profiles of a few lines with a Milne–Eddington model of the medium where the number of free parameters† is ten. More detailed models of the medium can be thought of, however. Consider, for instance, a model atmosphere like that of Eq. (9.1) where T, p_e, v_{LOS}, B, θ, and φ are specified in an optical depth grid of 40 points. The total number of free parameters of this model is 243 if constant ξ_{mic} and ξ_{mac} are assumed and a filling factor or stray-light parameter, f, is added. It is clear that we do not know which parameter and how much it should be modified in each step of the iterative procedure in order to approach the solution. Obviously, special strategies must be designed to deal with such a huge number of free parameters. Moreover, current instruments provide enormous amounts of data in very short times. Typically, several thousands or tens of thousands of such Stokes spectra are obtained in, say, half an hour. The use of automatic procedures is therefore mandatory.

11.1 The χ^2 merit function

As in many other branches of physics, let us assume that the Stokes profiles belong to $\mathcal{L}^2$, i.e., that they are square integrable.‡ The natural distance between, for example, the two Stokes profiles Q_1 and Q_2 is then their quadratic distance,

$$\int_{-\infty}^{\infty} |Q_1(\lambda) - Q_2(\lambda)|^2 \, d\lambda.$$

Our goal in the derivation of a physically meaningful model of the medium is to find a synthetic Stokes spectrum as close to the observed spectrum as possible. The model parameters of the synthetic spectrum will be the inferred parameters of the medium. Therefore, we shall resort to evaluating quadratic distances, as the reader may have already guessed.

Since we deal in practice with discrete samples of the Stokes profiles, consider a mean quadratic distance between the observed and the synthetic Stokes profiles given by a merit function such as

$$\chi^2(\boldsymbol{x}) \equiv \frac{1}{\nu} \sum_{s=0}^{3} \sum_{i=1}^{q} \left[I_s^{\text{obs}}(\lambda_i) - I_s^{\text{syn}}(\lambda_i; \boldsymbol{x}) \right]^2 w_{s,i}^2, \tag{11.1}$$

where s scans the four Stokes parameters and i the wavelength samples, and ν stands for the number of degrees of freedom, that is, the difference between the number of observables ($4q$) and that of the free parameters. Normalization by ν may be used as a "warning key" to remember that the number of free parameters

† Besides the nine free parameters mentioned in Section 9.4.1, actual Milne–Eddington techniques use another parameter that accounts for lack of spatial resolution or stray-light contamination of the observations.

‡ In fact, Stokes I do not belong to $\mathcal{L}^2$ but $I_c - I$ does. If the reader prefers, he or she can assume *local* square integrability of I, i.e., that the integrals exist over finite wavelength ranges.

cannot be unreasonably large; it permits a comparison among fits with different numbers of free parameters. The coefficients $w_{s,i}$ can be used in practice in order to weight some data more than others; for instance, they can be the inverse of measurement errors.

Equation (11.1) shows the χ^2 merit function as a scalar field in, for example, a 243-dimensional space! If we want to minimize the distance between observed and synthetic spectra, we face a formidable problem. Unlike Alice in Wonderland, we do care where we go and must therefore examine thoroughly all the possible paths and directions; in other words, *we need to know the derivatives of* χ^2 *with respect to the free parameters*. Inversion techniques are understood as those automated procedures bringing χ^2 to a physically meaningful minimum. Certainly, after the discussions of the present section, the reader may guess that inversion techniques may not simply be "automated trial-and-error techniques". Indeed, the specific technique we are discussing in this chapter is just one among several other possible (and conceivable) techniques for attacking the problem.

Before going into the mathematical details of the inversion technique, let us digress a little on the physical details of the problem. As anticipated in Chapter 10, the radiative transfer equation is the only available tool to succeed in modeling the medium. However, from a diagnostic point of view, it is unfortunate that the RTE is a differential equation with the Stokes spectrum as the unknown. The true unknowns for the observer are the free parameters of the model medium and they are included (through intricate non-linear dependences) in the RTE "coefficients": the propagation matrix and the source function vector. Thus, it is advisable *to invert the radiative transfer equation* and think of the formal solution, Eq. (9.20), as an integral equation where the true unknowns are in the integrand, and the Stokes spectrum does not play the role of an unknown but that of the observable. More specifically, Eq. (10.20), which explicitly gives the relationship between the modifications of the model parameters and the changes in the Stokes spectrum, contains all the necessary ingredients to move through the space of free parameters, as we shall see.

11.1.1 Derivatives of the χ^2 *merit function*

A perturbation of the Stokes parameters implies a modification of the χ^2 merit function given by

$$\delta\chi^2 = \frac{2}{\nu}\sum_{s=0}^{3}\sum_{i=1}^{q}\left[I_s^{\rm syn}(\lambda_i) - I_s^{\rm obs}(\lambda_i;\boldsymbol{x})\right] w_{s,i}^2\, \delta I_s^{\rm syn}(\lambda_i;\boldsymbol{x}), \tag{11.2}$$

where $\delta I_s^{\rm syn}(\lambda_i;\boldsymbol{x})$ is obtained from Eq. (10.20) or, numerically, from Eq. (10.21).

From a numerical point of view, considering the elements of the model atmosphere as depending on just one index is convenient since all the individual values of every physical quantity at each optical depth are considered as independent free parameters. Hence, we shall substitute $x_i(\tau_j)$ by x_p, where $p = (i-1)n + j$, n being the number of optical depth grid points. That is, p runs throughout the atmosphere for one physical quantity, then for another quantity, and so on. Certainly, p runs from 1 to $nm + r$, m being the number of physical quantities varying with depth and r that of the quantities which are constant (x_{m+i}, with $i = 1, 2, \ldots, r$). The same index contraction is useful for denoting response functions, so that we shall hereafter write $\boldsymbol{R}_p(\lambda_l)$ instead of

$$\Delta(\log\tau_c)\ln 10\, c_j \tau_j \boldsymbol{R}_i(\lambda_l;\tau_j)$$

or

$$\boldsymbol{R}'_{m+i}(\lambda_l),$$

the latter being the case for response functions to constant physical quantities.

With this simplified notation, Eq. (10.21) can be rewritten as

$$\delta \boldsymbol{I}^{\rm syn}(\lambda_l;\boldsymbol{x}) = \sum_{p=1}^{nm+r} \boldsymbol{R}_p(\lambda_l)\,\delta x_p, \tag{11.3}$$

which clearly emphasizes the role of response functions as partial derivatives of the Stokes spectrum with respect to the free parameters (which are assumed to be continuous variables). In fact, in a Milne–Eddington model, where $\boldsymbol{I}(0)$ has an analytic expression, $\boldsymbol{R}_p$ is substituted by $\partial\boldsymbol{I}(0)/\partial x_p$ that can be calculated analytically from Eq. (9.44).

After being introduced into Eq. (11.2), Eq. (11.3) gives

$$\delta\chi^2 = \sum_{p=1}^{nm+r}\left\{\frac{2}{\nu}\sum_{s=0}^{3}\sum_{i=1}^{q}\left[I_s^{\rm syn}(\lambda_i) - I_s^{\rm obs}(\lambda_i;\boldsymbol{x})\right] w_{s,i}^2\, R_{p,s}(\lambda_i)\right\}\delta x_p, \tag{11.4}$$

where the factor within braces is the direct partial derivative of the χ^2 merit function with respect to x_p:

$$\frac{\partial\chi^2}{\partial x_p} = \frac{2}{\nu}\sum_{s=0}^{3}\sum_{i=1}^{q}\left[I_s^{\rm syn}(\lambda_i) - I_s^{\rm obs}(\lambda_i;\boldsymbol{x})\right] w_{s,i}^2\, R_{p,s}(\lambda_i). \tag{11.5}$$

The second derivatives of χ^2 are given by

$$\frac{\partial^2\chi^2}{\partial x_p\partial x_k} = \frac{2}{\nu}\sum_{s=0}^{3}\sum_{i=1}^{q} w_{s,i}^2\left\{R_{p,s}(\lambda_i)\,R_{k,s}(\lambda_i) + \left[I_s^{\rm syn} - I_s^{\rm obs}\right]\frac{\partial R_{p,s}}{\partial x_k}\right\} \tag{11.6}$$

and, after neglecting the second term within braces, are approximated by

$$\frac{\partial^2 \chi^2}{\partial x_p \partial x_k} \simeq \frac{2}{\nu} \sum_{s=0}^{3} \sum_{i=1}^{q} w_{s,i}^2 \, [R_{p,s}(\lambda_i) \, R_{k,s}(\lambda_i)]. \tag{11.7}$$

The approximation is justified when either the derivative of $R_{p,s}$ is small compared to the first term or the difference in square brackets is small enough. The last case may be applicable when one is fairly near the minimum because that difference is expected to be close to the measurement error and hence uncorrelated with the model, so that the summation tends to cancel out the terms. As is understood in what follows, the second derivatives will be needed only when we are close to the minimum.†

11.2 The Marquardt method

Iterative procedures are the cornerstone of fitting data to models that depend non-linearly on the free parameters. Hence, we are dealing with an iteration whose steps allow one to move through the $(nm+r)$-dimensional space of the parameters in order to find a minimum of the χ^2 merit function. When one is close enough to the minimum, a parabolic expansion of χ^2 can be expected to apply, so that

$$\chi^2(\boldsymbol{x} + \delta\boldsymbol{x}) \simeq \chi^2(\boldsymbol{x}) + \delta\boldsymbol{x}^{\mathrm{T}} \, (\nabla\chi^2 + \mathbf{H}'\delta\boldsymbol{x}), \tag{11.8}$$

where, according to the general convention used in this book, a scalar product is understood in the second term on the right-hand side, where the elements of the gradient are given by Eq. (11.5), and where matrix $\mathbf{H}'$ is one half of the Hessian matrix, thus containing the second partial derivatives ($H'_{ij} = 1/2\, \partial^2\chi^2/\partial x_i \partial x_j$) made explicit in Eq. (11.7). If the second-order approximation is a good one, i.e., if we are very near the minimum of the merit function, then we can move to it by simply equating the term in parentheses in Eq. (11.8) to zero:

$$\nabla\chi^2 + \mathbf{H}'\delta\boldsymbol{x} = \boldsymbol{0}, \tag{11.9}$$

or, better,

$$\delta\boldsymbol{x} = -\mathbf{H}'^{-1} \nabla\chi^2. \tag{11.10}$$

If, however, we are far from the minimum and the approximation is not that good, the best we can do is to take a step down the gradient, that is,

$$\delta\boldsymbol{x} = a\nabla\chi^2, \tag{11.11}$$

† In any case, according to Press *et al.* (1986), this approximation ensures numerical stability.

where a must be a constant small enough not to lose the downhill path. The problem now is that we do not know the exact value of a, or at least its order of magnitude. The key of the Levenberg–Marquardt method (also called the Marquardt method) is the realization that the diagonal elements of the Hessian matrix provide such an idea of the value of a: note that if $\mathbf{H}'$ were diagonal, Eqs (11.10) and (11.11) would be formally identical:

$$\delta x_i = -\frac{1}{H'_{ii}} \nabla \chi^2.$$

But this scaling factor could still be too large, so the prescription is to divide it by a fudge factor, λ, which eventually can be set much greater than unity in order to reduce the step size. Therefore, if we keep just the gradient approximation, the right motion in the parameter space is given by

$$\delta x_i = -\frac{1}{\lambda H'_{ii}} \nabla \chi^2. \tag{11.12}$$

Now, very importantly, Eqs (11.10) and (11.12) can be combined into one equation,

$$\nabla \chi^2 + \mathbf{H}\, \delta \boldsymbol{x} = \boldsymbol{0}, \tag{11.13}$$

where the new matrix $\mathbf{H}$ is defined by

$$H_{ij} \equiv \begin{cases} H'_{ij}(1+\lambda), & \text{if } \ i = j, \\ H'_{ij}, & \text{if } \ i \neq j. \end{cases} \tag{11.14}$$

Therefore, when λ is very large, the modified Hessian matrix is forced to be quasi-diagonal so that the approximation is almost a first-order one [Eq. (11.12)]. If, however, λ becomes small, the second-order approximation applies. Then, by simply choosing different values of λ we change the applicability range of each step in the iteration. This feature is very helpful in practice. In fact, the following recipe is that recommended by the Marquardt method:

(1) Evaluate $\chi^2(\boldsymbol{x})$ with an initial guess at the model parameters.
(2) Take a modest value for λ, say $\lambda = 10^{-3}$.
(3) Solve Eq. (11.13) for $\delta \boldsymbol{x}$ and evaluate $\chi^2(\boldsymbol{x} + \delta \boldsymbol{x})$.
(4) If $\chi^2(\boldsymbol{x} + \delta \boldsymbol{x}) \geq \chi^2(\boldsymbol{x})$ we were too far away from the minimum. Therefore, increase λ significantly (by a factor of 10, say) and go back to step number 3.
(5) If $\chi^2(\boldsymbol{x} + \delta \boldsymbol{x}) \leq \chi^2(\boldsymbol{x})$ we were fairly close to the minimum, so that a decrease of λ (by a factor 10, for example) is recommended. Update the trial solution $\boldsymbol{x} + \delta \boldsymbol{x} \longrightarrow \boldsymbol{x}$ and go back to step number 3.

(6) To stop, wait until χ^2 decreases negligibly (1%, 0.1%, say) once or twice. Never stop when χ^2 increases since λ is readjusting itself to its optimum value.

11.2.1 Error calculation

Certainly, reaching the true minimum of the merit function is not guaranteed by the method. At the end of the procedure we will be hopefully close enough to the minimum so that the gradient term of Eq. (11.8) may be neglected and we may write

$$\Delta\chi^2 = \delta\boldsymbol{x}^{\mathrm{T}}\mathbf{H}'\delta\boldsymbol{x}, \tag{11.15}$$

where

$$\Delta\chi^2 \equiv \chi^2(\boldsymbol{x}_{\mathrm{min}}) - \chi^2(\boldsymbol{x}_0) \tag{11.16}$$

and

$$\delta\boldsymbol{x} \equiv \boldsymbol{x}_{\mathrm{min}} - \boldsymbol{x}_0, \tag{11.17}$$

$\boldsymbol{x}_{\mathrm{min}}$ and $\boldsymbol{x}_0$ being the model found by the numerical algorithm and that model where the true minimum of χ^2 is located, respectively. Since $\boldsymbol{x}_0$ has not been reached, the current value of $\Delta\chi^2$ can be produced statistically by several different models, $\boldsymbol{x}_{\mathrm{min}}$. Thus, the uncertainty of the inferred parameter x_k will be an average over all the possible realizations of the deviations δx_k, namely,†

$$\sigma_k^2 = \langle\delta x_k^2\rangle. \tag{11.18}$$

Since $\mathbf{H}'$ is real and symmetric, there always exists an orthogonal matrix $\mathbf{X}$ ($\mathbf{X}\mathbf{X}^{\mathrm{T}} = \mathbf{X}^{\mathrm{T}}\mathbf{X} = \mathbb{1}$) that diagonalizes it:

$$\boldsymbol{\Lambda} \equiv \mathbf{X}\mathbf{H}'\mathbf{X}^{\mathrm{T}}, \tag{11.19}$$

where $\boldsymbol{\Lambda} = \mathrm{diag}\,(\lambda_1, \lambda_2, \ldots, \lambda_{nm+r})$.

With this transformation, $\Delta\chi^2$ can be written as

$$\Delta\chi^2 = \boldsymbol{\pi}^{\mathrm{T}}\boldsymbol{\Lambda}\boldsymbol{\pi} = \sum_{p=1}^{nm+r} \lambda_p \pi_p^2, \tag{11.20}$$

where, obviously,

$$\boldsymbol{\pi} = \mathbf{X}\delta\boldsymbol{x}, \tag{11.21}$$

† All this section is based on Appendix B of a paper by Sánchez Almeida (1997).

so that

$$\delta x_k = \sum_{p=1}^{nm+r} X_{pk} \pi_p. \tag{11.22}$$

Hence, if $\mathbf{H}'$ is a good estimate of one half the "true" Hessian matrix (the Hessian matrix at $\boldsymbol{x}_0$), the uncertainty in x_k can be rewritten as

$$\sigma_k^2 = \sum_{l=1}^{nm+r} \sum_{p=1}^{nm+r} X_{lk} X_{pk} \langle \pi_l \pi_p \rangle \tag{11.23}$$

because matrix $\mathbf{X}$ is independent of the various realizations by hypothesis, and we just have to calculate the averages $\langle \pi_l \pi_p \rangle$. To do that, note that Eq. (11.20) is indeed the equation of a hypersphere in the $(nm + r)$-dimensional space of coordinates $\pi_p \sqrt{\lambda_p}$. Therefore, the averages are taken over its surface, whence $\langle \lambda_k \pi_k^2 \rangle$ must be the same for all $k = 1, 2, \ldots, nm + r$. In fact, Eq. (11.20) gives

$$\langle \pi_k^2 \rangle = \frac{\Delta \chi^2}{(nm + r)\lambda_k}. \tag{11.24}$$

Notice also that

$$\langle \pi_k \pi_l \rangle = 0 \quad \text{if} \quad k \neq l \tag{11.25}$$

because if Eq. (11.20) holds for $\boldsymbol{\pi}$, it also holds for $\boldsymbol{\pi}'$ such that all the components are equal to those of $\boldsymbol{\pi}$ except for one which is opposite. In summary,

$$\langle \pi_k \pi_l \rangle = \frac{\Delta \chi^2}{(nm + r)\lambda_k} \delta_{kl}, \tag{11.26}$$

where δ_{kl} is the Kronecker delta, whence

$$\sigma_k^2 = \frac{\Delta \chi^2}{nm + r} \sum_{p=1}^{nm+r} \frac{X_{pk}^2}{\lambda_p}. \tag{11.27}$$

The sum in Eq. (11.27) is easily identifiable as the k-th diagonal element of the inverse matrix of $\mathbf{H}'$. In fact, Eq. (11.19) implies that

$$\left[\mathbf{H}'^{-1}\right]_{ij} = \sum_{p=1}^{nm+r} \frac{X_{pi} X_{pj}}{\lambda_p}. \tag{11.28}$$

Therefore,

$$\sigma_k^2 = \frac{\Delta \chi^2}{nm + r} \left[\mathbf{H}'^{-1}\right]_{kk}. \tag{11.29}$$

An estimate of this uncertainty σ_p^2 of the inferred parameter x_p is given by†

$$\sigma_p^2 \simeq \frac{2}{nm+r} \frac{\sum_{s=0}^{3}\sum_{i=1}^{q}\left[I_s^{\rm syn}(\lambda_i) - I_s^{\rm obs}(\lambda_i)\right]^2 w_{s,i}^2}{\sum_{s=0}^{3}\sum_{i=1}^{q} R_{p,s}^2(\lambda_i)\, w_{s,i}^2}, \tag{11.30}$$

where $\Delta\chi^2$ has been approximated by $\chi^2(\boldsymbol{x}_{\rm min})$. The estimate is strictly valid only if $\mathbf{H}'$ is diagonal: we have substituted the inverse diagonal elements of $\mathbf{H}'$ for the diagonal elements of $\mathbf{H}'^{-1}$.

11.2.2 Problems in practice

Except for the adaptation of the gradient and Hessian to the expressions of Section 11.1.1 involving response functions, the Marquardt method, as explained so far, is completely general. Depending on the realization, however, one faces different practical problems. One cannot expect to find the same difficulties when dealing with a Milne–Eddington model of the free parameters as when dealing with a model that acknowledges the variation with optical depth of physical quantities. We are addressing our discussion to the latter case.‡

The two most important problems are related to the calculation of $\mathbf{H}^{-1}$. First, matrix $\mathbf{H}$ may be huge (243×243, for example) and inverting such enormous matrices is not an easy numerical task. Second, matrix $\mathbf{H}$ may be quasi-singular or numerically singular because of the different sensitivities of the Stokes spectrum to the various free parameters: we already know, for instance, that $\boldsymbol{I}$ is sensitive only to temperature in the very low layers of the atmosphere; hence the terms of the Hessian matrix involving, for example, the response function to perturbations of B at $\log\tau_{\rm c} \geq 0$, will be very approximately zero. To solve the first problem one can use successive approximation cycles in each of which the number of free parameters increases as the minimum of χ^2 is approached more and more. The first cycle works with a reasonable minimum number of parameters; then the number is increased for the second cycle, and so on. Besides, assumptions concerning the optical path dependence of the physical quantities of the model must be made. With regard to the second problem, the inversion of quasi-singular matrices is possible through the singular value decomposition (SVD) technique. However, a direct use of SVD is not convenient to our specific problem of radiative transfer and

† This estimate has already been used, and its reliability checked, in practice (e.g., Westendorp Plaza *et al.*, 2001).

‡ In fact, we are going to explain the special strategies designed for the so-called SIR technique of Ruiz Cobo and del Toro Iniesta (1992). SIR is the acronym for Stokes Inversion based on Response functions.

will be modified. The large difference in sensitivity of the Stokes spectrum to the various free parameters may result in some of them being neglected and their values forgotten during the inversion: the inclination angle, θ, of the magnetic field with respect to the line of sight may be less significant (its RF smaller) than the temperature, but we are also interested in inferring the θ values. Both the problems and the solutions will be better understood in the following two sections.

11.3 Parameters at the nodes and equivalent response functions

As mentioned above, the first problem we have to face is the large dimensions the Hessian matrix may have. There can be many ways of eliminating (or fixing) some of the free parameters. The reader is referred to the reviews cited in the recommended bibliography at the end of the chapter for finding the relevant literature on the subject. Let us concentrate here on the strategy proposed for the SIR technique that has been amply checked in practice with both synthetic and real data.

To understand the principles of the algorithm, assume that we are just dealing with one physical quantity varying with optical depth throughout the atmosphere ($m = 1; r = 0$). We shall reduce the number of free parameters by simply binding them all through an interpolation relationship. We shall assume that the perturbations, δx_k ($k = 1, 2, \ldots, n$), follow a given interpolation formula so that we need only deal with perturbations of a few depth grid points called *nodes*. Let n' be the number of such nodes. Obviously, $n' \leq n$, so that a reduction by a factor n'/n is effected over the Hessian matrix dimensions.

Let δy_l be the perturbation at the l-th node. Thus,

$$\delta y_l = \delta x_p \quad \text{with} \quad p = 1 + (l-1)\,\frac{(n-1)}{n'-1}, \tag{11.31}$$

where we can clearly appreciate that, by convention, the first and the last points of the grid ($p = 1$ and $p = n$) are nodes except when $n' = 1$ (when the above expression loses its meaning), in which case the perturbations are assumed to be constant throughout. Also, the relationship between p and l tells us that the possible number of nodes for a given grid is governed by the number of divisors of $n-1$. Assume that the perturbations at the various points are given by interpolation of the perturbations at the nodes:

$$\delta x_k = \sum_{l=1}^{n'} s_{k,l}\,\delta y_l, \tag{11.32}$$

where $s_{k,l}$ are coefficients that depend only on the ratio n/n'. Their specific values may be changed depending on the interpolation algorithm (e.g., cubic splines), but the interesting point in Eq. (11.32) is that we can express all the perturbations with the values of a few of them. Obviously, according to Eq. (11.31),

$$s_{p,l} = 1 \quad \text{when} \quad p = 1 + (l-1)\,\frac{(n-1)}{n'-1}. \tag{11.33}$$

The interpolation formula (11.32) governs the stratification with depth of the perturbations to a given physical quantity. Since the coefficients depend only on n/n', one may know the specific stratification by simply knowing n'. This is so because of the convention that the first and the last points of the grids are nodes. For example, within a given interpolation scheme, when $n' = 1$, constant perturbations are assumed that are taken to be equal to the mean of all perturbations; when $n' = 2$, the perturbations are supposed to follow a straight line passing through the perturbations at the nodes; when $n' = 3$, the perturbations vary parabolically with optical depth, and higher-order polynomials when $n' \geq 3$. It is important to remark that *the interpolation approximation constrains the possible stratifications of the perturbations, not those of the physical quantities.* If $n' = 1$, this means only that $(x_1, x_2, \ldots, x_n)$ will be modified during the whole procedure by constant perturbations, $\delta x_i = a$, $\forall\, i = 1, 2, \ldots, n$, to a final model $(x_1 + a, x_2 + a, \ldots, x_n + a)$, but if the original model were parabolic (or sinusoidal), the final model would also be parabolic (or sinusoidal). The variation with depth of the perturbations depends on the coefficients $s_{p,l}$, but the variation with depth of the model parameters is independent of these coefficients. Thus, *reducing the number of free parameters does not imply reducing the allowed complexity of the stratification of the physical quantities through the atmosphere.*

So far, we have dealt with just one physical quantity, but the concept of nodes and the interpolation approximation can certainly be extended to all the quantities relevant to line formation. The number, n', of nodes for temperature, for example, can indeed be different from that for the magnetic field strength or the line-of-sight velocity. In fact, those quantities assumed constant with optical depth, or which are single-valued, must necessarily have $n' = 1$ independently of the remainder. We shall assume here, however, that n' is the same for all the quantities that are functions of $\log \tau_{\mathrm{c}}$ in order not to complicate the notation. With this assumption, introducing Eq. (11.32) into Eq. (11.3), we get

$$\begin{aligned} \delta \boldsymbol{I}^{\mathrm{syn}}(\lambda_l; \boldsymbol{x}) &= \sum_{i=0}^{m-1} \sum_{p=1}^{n} \boldsymbol{R}_{in+p}(\lambda_l) \sum_{k=1}^{n'} s_{in+p,in'+k}\, \delta y_{in'+k} \\ &\quad + \sum_{p=1}^{r} \boldsymbol{R}_{nm+p}(\lambda_l)\, \delta y_{n'm+p}, \end{aligned} \tag{11.34}$$

where we have explicitly split index p in two in order to see better that the interpolation is carried out for every physical quantity that varies with optical depth. The second term corresponds to the constant quantities and, hence, each one is represented by one node. Note that, indeed, $s_{in+p,in'+k} = s_{p,k}$ since the interpolation coefficients are "periodic": in the first index with a "period" n and in the second index with a "period" n'. Changing the order of summation, Eq. (11.34) becomes

$$\begin{aligned} \delta \boldsymbol{I}^{\mathrm{syn}}(\lambda_l; \boldsymbol{x}) &= \sum_{i=0}^{m-1} \sum_{k=1}^{n'} \left\{ \sum_{p=1}^{n} s_{in+p,in'+k} \boldsymbol{R}_{in+p}(\lambda_l) \right\} \delta y_{in'+k} \\ &\quad + \sum_{k=1}^{r} \boldsymbol{R}_{nm+k}(\lambda_l)\, \delta y_{n'm+k}. \end{aligned} \tag{11.35}$$

Now, we can define the *equivalent response functions at the nodes* as

$$\tilde{\boldsymbol{R}}_{in'+k} \equiv \begin{cases} \displaystyle\sum_{p=1}^{n} s_{in+p,in'+k} \boldsymbol{R}_{in+p}(\lambda_l), & \text{if} \quad i < m, \\ \boldsymbol{R}_{ni+k}(\lambda_l), & \text{if} \quad i = m, \end{cases} \tag{11.36}$$

and with this definition we can finally write

$$\delta \boldsymbol{I}^{\mathrm{syn}}(\lambda_l; \boldsymbol{x}) = \sum_{p=1}^{n'm+r} \tilde{\boldsymbol{R}}_p(\lambda_l)\, \delta y_p. \tag{11.37}$$

The formal analogy between Eq. (11.37) and Eq. (11.3) is conspicuous so that we clearly understand our simplification of a problem with $nm + r$ free parameters in terms of another with $n'm + r$ free parameters. We can now follow all the steps from Section 11.1.1 through Section 11.2.1 after changing the starting equation. Note that the equivalent response functions at the nodes are indeed linear combinations of the response functions at all the points in the optical depth grid so that *we are in fact using information from the whole atmosphere, although weighted by the interpolation coefficients.* This is a very important and advantageous property. Since $\tilde{\boldsymbol{R}}_p$ is a linear combination of all the RFs throughout the atmosphere, parameters cannot be wildly varied because the whole stratification matters. Even those nodes with originally negligible response functions may now have sizable equivalent response functions: the shape of the corresponding quantity as a function of depth is somehow included in $\tilde{\boldsymbol{R}}_p$.

Let $\tilde{\mathbf{H}}$ be the new modified Hessian matrix, that is, a matrix built like $\mathbf{H}$ but from a Hessian matrix which contains the equivalent response functions. The Marquardt equation (11.13) then transforms to

$$\nabla \chi^2 + \tilde{\mathbf{H}}\, \delta \boldsymbol{y} = 0, \tag{11.38}$$

where we deal only with perturbations at the nodes.

The interpolation approximation is very useful for dividing the whole inversion procedure into separate cycles, each having a given number of nodes. In this way, a successive approximation can be performed if n' is increased from cycle to cycle. Although experience usually counsels a good starting guess model, it seems natural that this guess model may be quite far from the "true" one. If so, then a large n' might produce unwanted ripples in the stratification of the quantities: we may be too far from the χ^2 minimum so that errors in one free parameter are canceled out by errors in another. Therefore, when no previous knowledge of the actual atmosphere is available (as, for instance, during the inversion of a huge data set of spectropolarimetric observations), it is advisable to start with a cycle of iterations with $n' = 1$ or 2. Then, when the minimum with such conditions has been reached, a second cycle can be carried out with a larger n', and so on.

11.4 Less significant parameters and singular value decomposition

The second problem mentioned in Section 11.2.2 is related to the possible singularities of the modified Hessian matrix $\tilde{\mathbf{H}}$. Since we need to solve Eq. (11.38) for the perturbations, $\delta \boldsymbol{y}$, $\tilde{\mathbf{H}}$ has to be inverted:

$$\delta y_k = -\sum_{j=1}^{n'm+r} \left[\tilde{\mathbf{H}}^{-1}\right]_{kj} \frac{\partial \chi^2}{\partial y_j}. \tag{11.39}$$

If $\tilde{\mathbf{H}}$ is singular or numerically quasi-singular, the inversion cannot be performed. This is probably the case since we already know that the Stokes spectrum may be poorly sensitive to perturbations of certain physical quantities at certain optical depths. Some of those optical depths with small (equivalent) response functions are necessarily included in the numerical depth grid. Thus, several elements, and maybe some rows or columns, are probably close to zero or, even more probably, much smaller than others. As we shall see, this induces quasi-singularities of $\tilde{\mathbf{H}}$.

The numerical problem of inverting quasi-singular matrices is already solved with a number of techniques, among which the SVD technique provides the best solution in the least-squares sense (see Press *et al.* 1986), that is, it minimizes $|\nabla\chi^2 + \tilde{\mathbf{H}}\,\delta \boldsymbol{y}|^2$. In our particular case, since $\tilde{\mathbf{H}}$ is real and symmetric by construction, the SVD technique is based on its diagonalization.† There always exists an orthogonal matrix $\mathbf{Y}$ ($\mathbf{Y}^{\mathrm{T}}\mathbf{Y} = \mathbf{Y}\mathbf{Y}^{\mathrm{T}} = \mathbb{1}$) such that

$$\boldsymbol{\Gamma} \equiv \mathbf{Y}\tilde{\mathbf{H}}\mathbf{Y}^{\mathrm{T}}, \tag{11.40}$$

or

$$\tilde{\mathbf{H}} = \mathbf{Y}^{\mathrm{T}}\boldsymbol{\Gamma}\mathbf{Y}, \tag{11.41}$$

† The theorem of linear algebra underlying SVD is indeed more general and we do not discuss it here.

where $\mathbf{\Gamma} = \mathrm{diag}(\gamma_1, \gamma_2, \ldots, \gamma_{n'm+r})$ and γ_i are the eigenvalues of matrix $\tilde{\mathbf{H}}$, each associated with one free parameter. The inverse of the modified Hessian matrix is thus given by

$$\tilde{\mathbf{H}}^{-1} = \mathbf{Y}^{\mathrm{T}}\mathbf{\Gamma}^{-1}\mathbf{Y} \tag{11.42}$$

and certainly depends on the inverse of the $\tilde{\mathbf{H}}$ eigenvalues. If $\gamma_k = 0$, then its inverse goes to infinity and our matrix is singular. Besides strict singularity, the matrix is quasi-singular or *ill-conditioned* when the ratio between the largest and the lowest eigenvalue is so large that it approaches the machine's floating point precision.

The prescription of SVD for inverting $\tilde{\mathbf{H}}$ is just to replace $1/\gamma_k$ by zero, whenever γ_k is considered too small. Let, then, ϵ be a *tolerance* such that

$$\gamma_k \leq \epsilon \max\{\gamma_i, \ i = 1, 2, \ldots, n'm + r\} \longrightarrow \frac{1}{\gamma_k} = 0. \tag{11.43}$$

As a consequence of zeroing $1/\gamma_k$, the k-th free parameter does not contribute to the perturbations since, according to Eq. (11.42),

$$\left[\tilde{\mathbf{H}}^{-1}\right]_{ij} = \sum_{p=1}^{n'm+r} \frac{Y_{pi}Y_{pj}}{\gamma_p}. \tag{11.44}$$

Thus, if the tolerance is too large, the rank of $\tilde{\mathbf{H}}$ will be small, and the solution (11.39) will be fairly smooth. However, just a few free parameters will be affected. If ϵ is too small, "noisy" solutions are expected. In practice, ϵ takes values between 10^{-3} and 10^{-6}.

The neglecting of the smallest eigenvalues of matrix $\tilde{\mathbf{H}}$ in SVD introduces a new problem. In general, these small γ_k's are associated with those physical quantities whose perturbations produce small modifications in the Stokes spectrum, but although these quantities are perhaps less significant than others, it is nevertheless important that they be derived. $\boldsymbol{I}$ may be less sensitive to perturbations of θ or φ than to perturbations of B, but all three (B, θ, and φ) are needed to define the vector magnetic field. After the discussions in Chapter 10, where temperature was shown to be by far the most important quantity of line formation, the reader may understand that there exist some values of ϵ for which $\boldsymbol{B}$ or v_{LOS} might become unaffected through the whole inversion procedure. Although subtle on some occasions, the influence of the magnetic field or of the line-of-sight velocity cannot be neglected. Thus, a new strategy must be devised.

11.4.1 Modified singular value decomposition

If all these (quantitatively) less significant parameters have to be derived, one needs a way of discriminating among the various quantities and then among the values of each quantity at the different optical depths. The identification of the several free parameters is indeed not a difficult task. By construction of the vector model atmosphere, $\boldsymbol{y}$, we already know that

$$n_i \equiv \begin{cases} (i-1)n' + 1, & \text{if} \quad 1 \le i \le m, \\ (n'-1)m + i, & \text{if} \quad m < i \le m + r, \end{cases} \tag{11.45}$$

is the index of the first free parameter of the quantity number i. Note that the upper row of the definition (11.45) corresponds to physical quantities which vary with optical depth; the lower row corresponds to constant or single-valued quantities.

Therefore, it is possible to decompose $\delta\boldsymbol{y}$ as a sum,

$$\delta\boldsymbol{y} = \sum_{i=1}^{m+r} \delta\boldsymbol{y}^i, \tag{11.46}$$

of $m + r$ vectors of $n'm + r$ components each but with zeroes everywhere except in those elements corresponding to a given quantity:

$$\delta y^i_j \equiv \begin{cases} \delta y_j, & \text{if} \quad n_i \le j < n_{i+1}, \\ 0, & \text{otherwise.} \end{cases} \tag{11.47}$$

Let us now decompose matrix $\tilde{\mathbf{H}}$ as a sum of matrices such that each acts on one and only one $\delta\boldsymbol{y}^i$. To do that, define $m + r$ matrices $\mathbf{Y}^i$ such that

$$Y^i_{jk} \equiv \begin{cases} Y_{jk}, & \text{if} \quad n_i \le j < n_{i+1}, \\ 0, & \text{otherwise.} \end{cases} \tag{11.48}$$

With this definition,

$$\mathbf{Y} = \sum_{i=1}^{m+r} \mathbf{Y}^i \tag{11.49}$$

and

$$\mathbf{Y}^{i^{\mathrm{T}}} \delta\boldsymbol{y} = \mathbf{Y}^{\mathrm{T}} \delta\boldsymbol{y}^i. \tag{11.50}$$

Let $\boldsymbol{\Gamma}^i$ be defined as

$$\boldsymbol{\Gamma}^i \equiv \boldsymbol{\Gamma} \mathbf{Y}^{i^{\mathrm{T}}} \mathbf{Y}, \tag{11.51}$$

so that, trivially,

$$\sum_{i=1}^{m+r} \mathbf{\Gamma}^i = \mathbf{\Gamma}. \tag{11.52}$$

Matrix $\mathbf{\Gamma}^i$ turns out to be that component of $\mathbf{\Gamma}$ which acts only on δy^i because, if we call

$$\tilde{\mathbf{H}}^i \equiv \mathbf{Y}\mathbf{\Gamma}^i\mathbf{Y}^{\mathrm{T}}, \tag{11.53}$$

we then have from Eq. (11.51) that

$$\tilde{\mathbf{H}}^i\, \delta y = \mathbf{Y}\mathbf{\Gamma}\mathbf{Y}^{i\,\mathrm{T}}\mathbf{Y}\mathbf{Y}^{\mathrm{T}}\delta y; \tag{11.54}$$

from the orthogonality of matrix $\mathbf{Y}$,

$$\tilde{\mathbf{H}}^i\, \delta y = \mathbf{Y}\mathbf{\Gamma}\mathbf{Y}^{i\,\mathrm{T}}\delta y, \tag{11.55}$$

and from Eq. (11.50),

$$\tilde{\mathbf{H}}^i\, \delta y = \tilde{\mathbf{H}}\, \delta y^i. \tag{11.56}$$

We have thus isolated the action of the modified Hessian matrix on the various physical quantities. Therefore, we can manipulate every matrix $\mathbf{\Gamma}^i$ as one does normally through the regular SVD technique. Those diagonal matrix elements that are smaller than the tolerance multiplied by the maximum diagonal element will be zeroed:

$$\Gamma^i_{jj} \le \epsilon \max\{\Gamma^i_{kk}\} \longrightarrow \Gamma^i_{jj} = 0. \tag{11.57}$$

Since, according to Eq. (11.52), the eigenvalues of $\tilde{\mathbf{H}}$ are sums of Γ^i_{jj}'s,

$$\gamma_k = \sum_{i=1}^{m+r} \Gamma^i_{kk}, \tag{11.58}$$

when inverting $\tilde{\mathbf{H}}$ we shall proceed by zeroing those $1/\gamma_k$'s whose $\gamma_k = 0$:

$$\gamma_k = 0 \longrightarrow \frac{1}{\gamma_k} = 0. \tag{11.59}$$

With this modified singular value decomposition technique, we ensure that at least one free parameter (one node) of every physical quantity is taken into account in each iteration, so that it will be modified according to its significance, that is, to the relative weight of its equivalent response function.

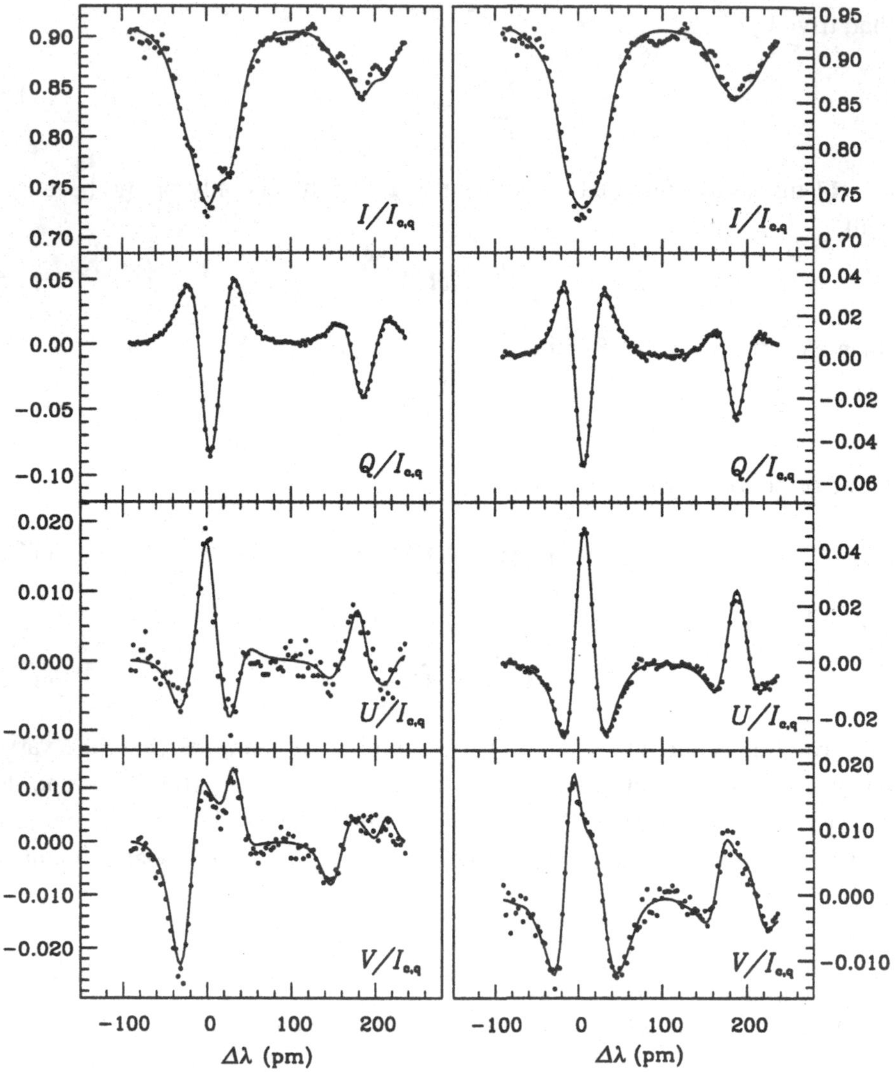

Fig. 11.2. Stokes profiles of three infrared Fe I lines at 1558.83, 1559.01, and 1559.07 nm (the latter two are blended) as observed with TIP on two penumbral points of a sunspot. Dots are the observations and the fit with SIR is shown in solid lines. From del Toro Iniesta *et al.* (2001).

11.5 An example

This section is devoted to giving just a flavor of the use of SIR for analyzing real spectropolarimetric data. Of the many examples that could have been given, only one has been selected because it somehow points to the future. Figures 11.2 and 11.3 show the results of SIR for two points in the penumbra of a sunspot as observed with TIP (see Section 5.3.3). Three Fe I lines at 1558.83, 1559.01, and

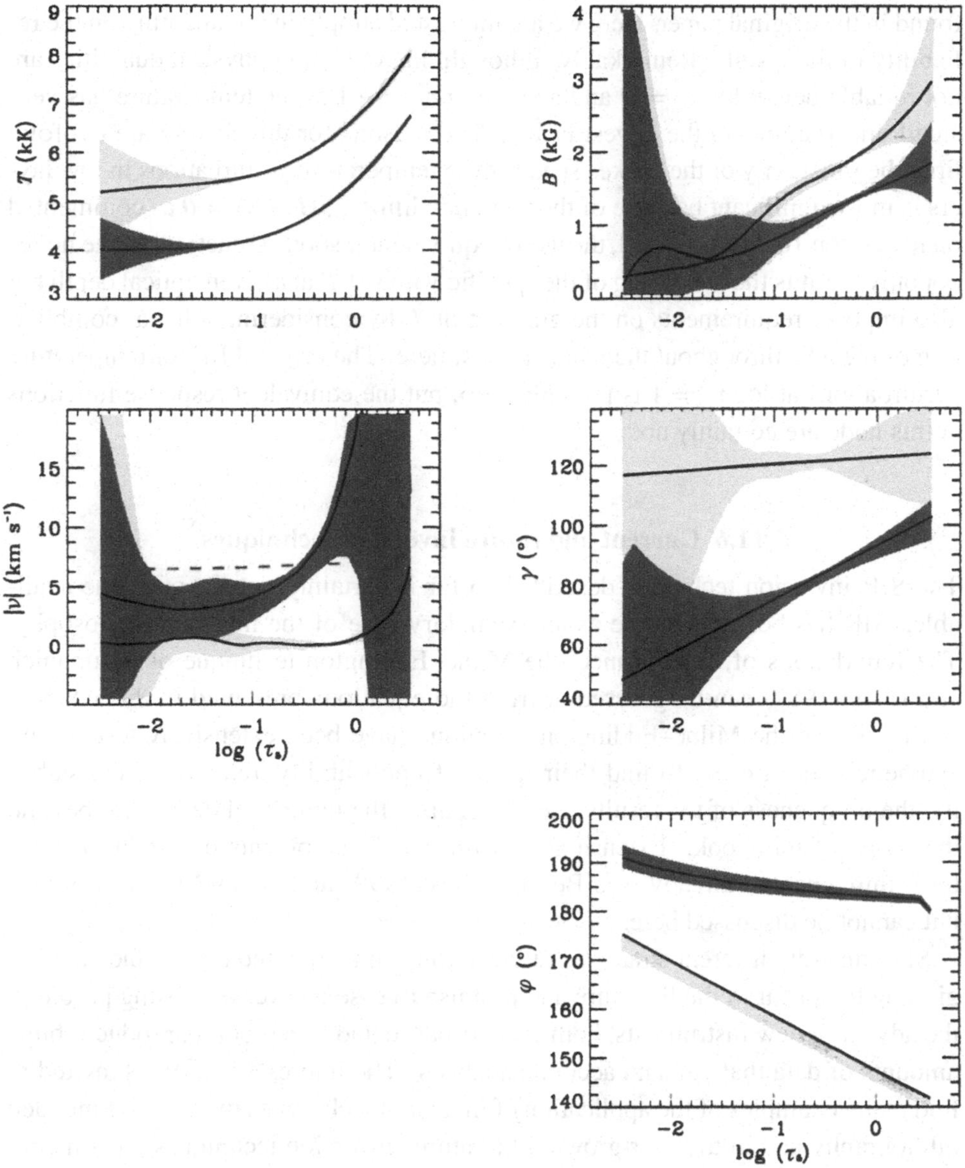

Fig. 11.3. Two-component model atmosphere which produces the fitted Stokes profiles on the left column of Figure 11.2. The model physical quantities are represented by solid lines. Uncertainty ranges are shown as shadings. The dashed line of the velocity panel indicates the local speed of sound. From del Toro Iniesta *et al.* (2001).

1559.07 nm are observed and inverted (the latter two are blended). In these particular two cases, the model atmosphere consists of two components, each with its own stratification of the physical quantities and discernible in Fig. 11.3 with a different shading of the uncertainties. The astronomical implications of the results are to be found in the original paper. Here we are interested simply in commenting on the reliability of the results. Remarkably, although the remaining physical quantities are not reliable below $\log \tau_c = 0$ and above $\log \tau_c = -1.5$, the temperature has very small uncertainties in the lowest layers. The reasons for this feature are twofold: first, the sensitivity of the Stokes spectrum to temperature perturbations in low layers is most significant because of the tendency $\lim_{\tau_c \to \infty}(\boldsymbol{I} - \boldsymbol{S}) = \boldsymbol{0}$ as commented on in Section 10.3.2.2; second, the use of equivalent response functions at the nodes not only permits the weighting of the specific value of T at a given optical depth but also imposes requirements on the gradient of T by considering a linear combination of the RFs throughout the whole atmosphere. The original RF for temperature perturbations at $\log \tau_c = 1$ is possibly zero, but the equivalent response functions at this node are certainly not.

11.6 Current and future inversion techniques

The SIR inversion technique described so far is certainly not the only one available. SIR has been used here as an exemplary case of the inversion philosophy. The foundations of, for instance, the Milne–Eddington technique of Skumanich and Lites (1987) can be guessed at from the equations presented in this chapter. Both SIR and the Milne–Eddington technique have been extensively tested with synthetic data in order to find their range of applicability, reliability, and stability, the uniqueness of the results, etc. Of course, the numerical details are beyond the scope of this book. Extensions and further developments of SIR itself have been implemented already [see Bellot Rubio (1998) and Socas Navarro (1999)], but cannot be discussed here.

Significantly different strategies of inversion can be devised and, in fact, are beginning to appear in the literature and promise to ease an ever increasing problem: the advent of new instruments, both ground-based and space-borne, produces huge amounts of data that demand accurate analysis. The interested reader is invited to find many examples of the application of inversion techniques in the recommended bibliography and in the ever-growing literature: inversion techniques are a matter of continuous research and development.

Recommended bibliography

Allende Prieto, C. (1999). *Surface inhomogeneities and semi-empirical modeling of metal-poor stellar photospheres.* PhD Thesis (Universidad de La Laguna: La Laguna, Spain). Chapter 2.

Bellot Rubio, L.R. (1998). *Structure of solar magnetic elements from the inversion of Stokes spectra.* PhD Thesis (Universidad de La Laguna: La Laguna, Spain). Chapter 6.

Press, W.H., Flannery, B.P., Teukolsky, S.W., and Vetterling, W.T. (1986). *Numerical recipes*. (Cambridge University Press: Cambridge). Chapters 2 and 14. Sections 2.9, 14.4 and 14.5.

Rees, D., López Ariste, A., Thatcher, J., and Semel, M. (2000). Fast inversion of spectral lines using principal component analysis. I. Fundamentals. *Astron. Astrophys.* **355**, 759.

Ruiz Cobo, B. (1992). *Inversión de la ecuación de transporte radiativo*. PhD Thesis (Universidad de La Laguna: La Laguna, Spain). Chapters 4 and 5.

Ruiz Cobo, B. and del Toro Iniesta, J.C. (1992). Inversion of Stokes profiles. *Astrophys. J.* **398**, 385.

Sánchez Almeida, J. (1997). Physical properties of the solar magnetic photosphere under the MISMA hypothesis. I. Description of the inversion procedure. *Astron. Astrophys.* **491**, 993.

Skumanich, A. and Lites, B.W. (1987). Stokes profile analysis and vector magnetic fields. I. Inversion of photospheric lines. *Astrophys. J.* **322**, 473.

Socas Navarro, H. (1999). *NLTE Inversion of spectral lines and Stokes profiles*. PhD Thesis (Universidad de La Laguna: La Laguna, Spain). Chapter 3.

Socas-Navarro, H. and López Ariste, A. (2001). Fast inversion of spectral lines using principal component analysis. II. Inversion of real Stokes data. *Astrophys. J.* **553**, 949.

del Toro Iniesta, J.C. (2001). Solar polarimetry and magnetic field measurements, in *The dynamic sun*. A. Hanslmeier, M. Messerotti, and A. Veronig (eds.) (Kluwer Academic Pub.: Dordrecht), p. 183.

del Toro Iniesta, J.C., Bellot Rubio, L.R., and Collados, M. (2001). Cold, supersonic Evershed downflow in a sunspot. *Astrophys. J.* **549**, L139.

del Toro Iniesta, J.C. and Ruiz Cobo, B. (1995). Calibration of magnetic fields from spectropolarimetric measurements, in *La polarimétrie, outil pour l'étude de l'activité magnétique solaire et stellaire*. N. Mein and S. Sahal-Bréchot (eds.) (Observatoire de Paris: Paris), p. 127.

del Toro Iniesta, J.C. and Ruiz Cobo, B. (1996). Stokes profiles inversion techniques. *Solar Phys.* **164**, 169.

del Toro Iniesta, J.C. and Ruiz Cobo, B. (1997). Inversion of Stokes profiles: what's next?, in *Forum THEMIS. Science with THEMIS*. N. Mein and S. Sahal-Bréchot (eds.) (Observatoire de Paris: Paris), p. 93.

del Toro Iniesta, J.C., Tarbell, T.D., and Ruiz Cobo, B. (1994) On the temperature and velocity through the photosphere of a sunspot penumbra, *Astrophys. J.* **436**, 400.

Westendorp Plaza, C., del Toro Iniesta, J.C., Ruiz Cobo, B., Martínez Pillet, V., Lites, B.W., and Skumanich, A. (1998). Optical tomography of a sunspot. I. Comparison between two inversion techniques. *Astrophys. J.* **494**, 453.

Westendorp Plaza, C., del Toro Iniesta, J.C., Ruiz Cobo, B., Martínez Pillet, V., Lites, B.W., and Skumanich, A. (2001). Optical tomography of a sunspot. II. Vector magnetic field and temperature stratification. *Astrophys. J.* **547**, 1130.

Westendorp Plaza, C., del Toro Iniesta, J.C., Ruiz Cobo, B., and Martínez Pillet, V. (2001). Optical tomography of a sunspot. III. Velocity stratification and the Evershed effect. *Astrophys. J.* **547**, 1148.

Index